STRUCTURAL HEALTH MONITORING - MEASUREMENT METHODS AND PRACTICAL APPLICATIONS

Edited by **Moises Rivas-Lopez, Wendy Flores Fuentes** and **Oleg Sergiyenko**

Structural Health Monitoring - Measurement Methods and Practical Applications
http://dx.doi.org/10.5772/65818
Edited by Moises Rivas-Lopez, Wendy Flores Fuentes and Oleg Sergiyenko

Contributors

Shun Weng, Hongping Zhu, Yong Xia, Fei Gao, Shaofei Jiang, Shenglan Ma, Yanhua Sun, Shiwei Liu, Wenjia Ma, Tat Hean Gan, Ángela Angulo, Slim Soua, Graham Edwards, Feng-Liang Zhang, Yan-Chun Ni, Ponciano Escamilla-Ambrosio, X Liu, J M Ramirez-Cortes, Abraham Rodríguez-Mota, Pilar Gómez-Gil

Published by InTech
Janeza Trdine 9, 51000 Rijeka, Croatia

Publishing Process Manager Mirena Calmic
Technical Editor SPi Global
Cover InTech Design team

First published June, 2017
Printed in Croatia
Legal deposit, Croatia: National and University Library in Zagreb

Additional hard copies can be obtained from orders@intechopen.com

Structural Health Monitoring - Measurement Methods and Practical Applications, Edited by Moises Rivas-Lopez, Wendy Flores Fuentes and Oleg Sergiyenko
p. cm.
Print ISBN 978-953-51-3253-0
Online ISBN 978-953-51-3254-7

For EU product safety concerns:
IN TECH d.o.o., Prolaz Marije Krucifikse Kozulić 3, 51000 Rijeka, Croatia,
info@intechopen.com or visit our website at intechopen.com.

Meet the editors

Dr. Rivas-López Moisés was born in 1960. He received the BS and MS degrees in Autonomous University of Baja California, Mexico, in 1985 and 1991, respectively, and the PhD degree in Science, Applied Physics, in the same University, in 2010. He has written 5 book chapters and 35 Journal and Proceedings Conference papers in optoelectronics and control applications. Also, he has presented different works in several International Congresses of IEEE, ICROS, SICE, in America and Europe. Dr. Rivas was Dean of Engineering Institute of Autonomous University Baja California (1997–2005) and Rector of Polytechnic University of Baja California (2006–201). He is a member of National Researcher System and now is the head of Physics Engineering Department, of Engineering Institute of UABC, Mexico.

Dr. Flores Fuentes was born in Baja California, Mexico on January, 1978. She received the bachelor's degree in Electronic Engineering from the Autonomous University of Baja California in 2001, the master's in Engineering degree from Technological Institute of Mexicali in 2006, and the PhD. degree in Science, Applied Physics, from Autonomous University of Baja California in June 2014. Until now she has authored 4 journal articles and 2 books with various publishers, and 13 proceedings articles in IEEE events. Recently she organized and participated as Chair of Special Session on "Machine Vision, Control and Navigation" at IEEE ISIE 2015. She has been incorporated to CONACYT National Research System in 2016.

Oleg Sergiyenko received his BS and MS degrees from Kharkiv National University of Automobiles and Highways, Kharkiv, Ukraine, in 1991 and 1993, respectively. He received his PhD degree from Kharkiv National Polytechnic University in 1997. He has written 81 papers in control systems, robot navigation, 3D coordinate measurement, and SHM. He is an editor of two books, holds two patents in Ukraine and Mexico, and is a reviewer for various publishers. He participated as a reviewer and session chair in several IEEE conferences in different countries and holds several "Best Presentation Awards." From December 2004 to present, he is a full-time researcher and head of Applied Physics Department in Engineering Institute of Baja California Autonomous University, Mexico.

Contents

Preface IX

Chapter 1 **Substructuring Method in Structural Health Monitoring 1**
Shun Weng, Hong-Ping Zhu, Yong Xia and Fei Gao

Chapter 2 **A New Method of SHM for Steel Wire Rope and its Apparatus 23**
Shiwei Liu, Yanhua Sun and Wenjia Ma

Chapter 3 **Structural Damage Detection Based on Improved Multi-Particle Swarm Co-Evolution Optimization Algorithm 47**
Shaofei Jiang and Shenglan Ma

Chapter 4 **Operational Modal Analysis of Super Tall Buildings by a Bayesian Approach 65**
Feng-Liang Zhang and Yan-Chun Ni

Chapter 5 **Mooring Integrity Management: Novel Approaches Towards In Situ Monitoring 87**
Ángela Angulo, Graham Edwards, Slim Soua and Tat-Hean Gan

Chapter 6 **Multi-Sensor Feature Extraction and Data Fusion Using ANFIS and 2D Wavelet Transform in Structural Health Monitoring 109**
Ponciano Jorge Escamilla-Ambrosio, Xuefeng Liu, Juan Manuel Ramírez-Cortés, Abraham Rodríguez-Mota and María del Pilar Gómez-Gil

Preface

The continuous and rapid growth of industrial and service infrastructure in all countries of the world has given rise to a new field of engineering, called *structural health monitoring* (SHM), which basically deals with performance and damage detection in man-made structures and space vehicles and monitoring of geological faults.

Damage detection starts with the acquisition of data, obtained from sensors and with monitoring of systems. These data are used to model the behavior of structures under adverse scenarios, in order to find possible anomalies.

Throughout SHM history, many systems for damage detection have been used with sensors based on different technologies like optical fiber, video cameras, passive and active optical scanners, wireless networks, and piezoelectric transducers, among others. Each of these systems has advantages and disadvantages regarding the type of structure and variables to monitor, as well as the kind of potential damage that the structure could suffer and must be prevented, to preserve infrastructure and prevent loss of human lives.

The importance of SHM lies in the fact that these systems can be used to monitor the health of a structure using a nondestructive measuring method.

The present book includes six chapters with theoretical models and relevant examples of practical applications. Chapter 1 provides a forward and inverse substructuring method for model updating of large-scale structures. Chapter 2 presents an improved multiparticle swarm coevolution optimization algorithm for damage detection. Chapter 3 deals primarily with ambient modal identification of four super tall buildings using a Bayesian approach. In Chapter 4, a mooring integrity management for inspection in situ of chain monitoring is presented. Chapter 5 illustrates a solid research activity in an interesting and novel method for SHM of steel wire ropes, and Chapter 6 proposes a novel approach in which ANFIS and 2D WT technologies were combined to perform structural damage identification.

Each chapter has been written by specialists in the area and includes a complete background using numerous references of actual and novel research articles.

The objective of this book is to be considered as a research reference, practical textbook, or supplement material in graduate programs.

Finally, we are very thankful to the authors for their kind contributions and also appreciate the publisher support and guidance to publish this book.

Moises Rivas López, Wendy Flores Fuentes and **Oleg Sergiyenko**
Engineering Institute
Autonomous University of Baja California
Mexico

Chapter 1

Substructuring Method in Structural Health Monitoring

Shun Weng, Hong-Ping Zhu, Yong Xia and Fei Gao

Additional information is available at the end of the chapter

http://dx.doi.org/10.5772/67890

Abstract

In sensitivity-based finite element model updating, the eigensolutions and eigensensitivities are calculated repeatedly, which is a time-consuming process for large-scale structures. In this chapter, a forward substructuring method and an inverse substructuring method are proposed to fulfill the model updating of large-scale structures. In the forward substructuring method, the analytical FE model of the global structure is divided into several independent substructures. The eigensolutions of each independent substructure are used to recover the eigensolutions and eigensensitivities of the global structure. Consequently, only some specific substructures are reanalyzed in model updating and assembled with other untouched substructures to recover the eigensolutions and eigensensitivities of the global structure. In the inverse substructuring method, the experimental modal data of the global structure are disassembled into substructural flexibility. Afterwards, each substructure is treated as an independent structure to reproduce its flexibility through a model-updating process. Employing the substructuring method, the model updating of a substructure can be conducted by measuring the local area of the concerned substructure solely. Finally, application of the proposed methods to a laboratory tested frame structure reveals that the forward and inverse substructuring methods are effective in model updating and damage identification.

Keywords: structural health monitoring, substructuring method, damage identification, eigensolutions, eigensensitivity

1. Introduction

Accurate finite element (FE) models are essential in damage identification and condition assessment for structural health monitoring. In vibration-based model-updating process, the FE model of a structure is iteratively updated to guarantee its vibration properties to reproduce the measured counterparts in an optimal manner [1]. In the optimization process, the structural responses are usually used to construct the objective function. The response sensitivities, which are the first derivatives of the structural responses to some structural physical parameters, are used to indicate a rapid searching direction. In this regard, the eigensolutions and their associated sensitivity matrices of the analytical model are required to be gained repeatedly in each iteration [2, 3]. The majority of the practical structures in civil engineering are large in scale, thus their FE models usually consists of a large number of degrees of freedom (DOFs) and uncertain updating parameters. The conventional model updating methods of large-scale structures are expensive in terms of computation time and computer memory [2].

It has been proved that the substructuring methods are efficient in dealing with large-scale structures, as it takes the local area as an independent structure [4–9]. First, the global structure is divided rationally into several smaller substructures to make it much easier and faster to analyze the small substructures independently. Second, the FE model of a substructure has much fewer uncertain parameters than the global structure, which helps to accelerate the convergence of optimization process to identify these parameters and alleviates the ill-condition problems. Third, the substructuring method is required to measure the local area of the practical structure and save the experimental instruments. Finally, the substructuring method can be more promising if combined with parallel computation.

In this chapter, a forward substructuring method and an inverse substructuring method are proposed for model updating and damage identification. In the forward substructuring method, the divided substructures are analyzed independently and are assembled to recover the eigensolutions of the global structure by satisfying the coordination condition of displacement at the interfaces. Afterwards, the fast-calculated eigensolutions and eigensensitivities of the global structure are used for model updating. In the inverse substructuring method, the experimental modal data of the global structure are disassembled into the substructural flexibility by satisfying the coordination condition of force and displacement at the interfaces. Based on the extracted substructural flexibility, the model-updating process is performed on the concerned substructure by treating it as an independent structure. In the following part, the forward and inverse substructuring methods will be explained first and then the two kinds of substructure-based model updating methods will be verified by a laboratory-tested frame structure.

2. Forward substructuring method

2.1. Eigensolutions

In the forward substructuring method, the eigensolutions and eigensensitivities of a substructure are calculated and assembled to recover those of the global structure. The global structure http://dx.doi.org/10.5772/67890

is divided into N_S independent substructures, and the number of DOFs of each substructure is n_j (j = 1,2,…, N_S). Treated as an independent structure, the eigenequation of the jth substructure is expressed as

$$\mathbf{K}^{(j)}\{\phi_i^{(j)}\} = \lambda_i^{(j)}\mathbf{M}^{(j)}\{\phi_i^{(j)}\} \tag{1}$$

where **K**(j) and **M**(j) are the stiffness matrix and mass matrix of the jth substructure, respectively. ($\phi i(j)$, $\lambda i(j)$) are the ith eigenpairs of the jth substructure. The $n^{(j)}$ pairs of eigenvalues and eigenvectors are expressed as [10]

$$\mathbf{\Lambda}^{(j)} = Diag\left[\lambda_1^{(j)}, \lambda_2^{(j)}, \ldots, \lambda_{n_j}^{(j)}\right], \mathbf{\Phi}^{(j)} = \left[\phi_1^{(j)}, \phi_2^{(j)}, \ldots, \phi_{n_j}^{(j)}\right],$$

And due to orthogonality, eigenvectors satisfy the two following formulas as

$$\left[\mathbf{\Phi}^{(j)}\right]^T \mathbf{K}^{(j)}\mathbf{\Phi}^{(j)} = \mathbf{\Lambda}^{(j)}, \left[\mathbf{\Phi}^{(j)}\right]^T \mathbf{M}^{(j)}\mathbf{\Phi}^{(j)} = n_j$$

The eigensolutions of the global structure can be recovered by adding constraints at the interfaces to obey the principle of virtual work and geometric compatibility like [11]

$$\begin{bmatrix} \mathbf{\Lambda}^p - \overline{\lambda}\mathbf{I} & -\mathbf{\Gamma} \\ -\mathbf{\Gamma}^T & \mathbf{0} \end{bmatrix} \begin{Bmatrix} \mathbf{z} \\ \tau \end{Bmatrix} = \begin{Bmatrix} \mathbf{0} \\ \mathbf{0} \end{Bmatrix} \tag{2}$$

where

$$\begin{aligned} &\mathbf{\Gamma} = [\mathbf{C}\mathbf{\Phi}^p]^T, \mathbf{\Lambda}^p = Diag\left[\mathbf{\Lambda}^{(1)}, \mathbf{\Lambda}^{(2)}, \ldots, \mathbf{\Lambda}^{(N_s)}\right] \\ &\mathbf{\Phi}^p = Diag\left[\mathbf{\Phi}^{(1)}, \mathbf{\Phi}^{(2)}, \ldots, \mathbf{\Phi}^{(N_s)}\right] \end{aligned} \tag{3}$$

Matrix **C** gives the general implicit constraints to guarantee the nodes at the interface identical displacement [11]. **C** contains two nonzero elements in each row, which are 1 and −1 for a rigid interface connection. $\mathbf{\Lambda}^p$ and $\mathbf{\Phi}^p$ are diagonally assembled from the eigensolutions of each substructure. $\overline{\lambda}$ is the eigenvalue of the global structure, which is the square of circular frequencies. The eigenvectors of the global structure are recovered by $\overline{\mathbf{\Phi}} = \mathbf{\Phi}^p\{\mathbf{z}\}$. τ indicates the interface forces between the adjacent substructures. Superscript "p" denotes the primitive matrices, which is assembled diagonally from the substructural matrices before displacement constraints at the adjacent substructures are imposed.

It is noted from Eq. (2) that $\mathbf{\Lambda}^p$ and $\mathbf{\Phi}^p$ are assembled from all modes of the substructures. It is inefficient and unworthy with all eigenmodes available, as only the first few eigenmodes are usually required for a large-scale structure. Here, the first few eigensolutions of each substructure are selected as "master" modes, and the residual higher modes are the "slave" modes. Only the master modes are used to gain the eigenequation of the global structure.

From here on, subscript "m" represents the "master" modes and subscript "s" denotes the "slave" modes, respectively. The eigenequation (Eq. (2)) is then rewritten according to the master modes and slave modes as

$$\begin{bmatrix} \mathbf{\Lambda}_m^p - \overline{\lambda}\mathbf{I} & \mathbf{0} & -\mathbf{\Gamma}_m \\ \mathbf{0} & \mathbf{\Lambda}_s^p - \overline{\lambda}\mathbf{I} & -\mathbf{\Gamma}_s \\ -\mathbf{\Gamma}_m^T & -\mathbf{\Gamma}_s^T & \mathbf{0} \end{bmatrix} \begin{Bmatrix} \mathbf{z}_m \\ \mathbf{z}_s \\ \tau \end{Bmatrix} = \begin{Bmatrix} \mathbf{0} \\ \mathbf{0} \\ \mathbf{0} \end{Bmatrix} \tag{4}$$

where

$$\begin{aligned} &\mathbf{\Lambda}_m^p = Diag[\mathbf{\Lambda}_m^{(1)}, \mathbf{\Lambda}_m^{(2)}, \ldots, \mathbf{\Lambda}_m^{(j)}, \ldots, \mathbf{\Lambda}_m^{(N_s)}], \mathbf{\Lambda}_m^{(j)} = Diag[\lambda_1^{(j)}, \lambda_2^{(j)}, \ldots, \lambda_{m^{(j)}}^{(j)}] \\ &\mathbf{\Phi}_m^p = Diag[\mathbf{\Phi}_m^{(1)}, \mathbf{\Phi}_m^{(2)}, \ldots, \mathbf{\Phi}_m^{(j)}, \ldots, \mathbf{\Phi}_m^{(N_s)}], \mathbf{\Phi}_m^{(j)} = [\phi_1^{(j)}, \phi_2^{(j)}, \ldots, \phi_{m^{(j)}}^{(j)}] \\ &\mathbf{\Lambda}_s^p = Diag[\mathbf{\Lambda}_s^{(1)}, \mathbf{\Lambda}_s^{(2)}, \ldots, \mathbf{\Lambda}_s^{(j)}, \ldots, \mathbf{\Lambda}_s^{(N_s)}], \mathbf{\Lambda}_s^{(j)} = Diag[\lambda_{m^{(j)}+1}^{(j)}, \lambda_{m^{(j)}+2}^{(j)}, \ldots, \lambda_{m^{(j)}+s^{(j)}}^{(j)}] \\ &\mathbf{\Phi}_s^p = Diag[\mathbf{\Phi}_s^{(1)}, \mathbf{\Phi}_s^{(2)}, \ldots, \mathbf{\Phi}_s^{(j)}, \ldots, \mathbf{\Phi}_s^{(N_s)}], \mathbf{\Phi}_s^{(j)} = [\phi_{m^{(j)}+1}^{(j)}, \phi_{m^{(j)}+2}^{(j)}, \ldots, \phi_{m^{(j)}+s^{(j)}}^{(j)}] \\ &\mathbf{\Gamma}_m = [\mathbf{C}\mathbf{\Phi}_m^p]^T, \mathbf{\Gamma}_s = [\mathbf{C}\mathbf{\Phi}_s^p]^T \\ &m^p = \sum_{j=1}^{N_s} m_j, s^p = \sum_{j=1}^{N_s} s_j, m_j + s_j = n_j (j = 1, 2, \ldots, N_s) \end{aligned} \tag{5}$$

According to the second line of Eq. (4), the slave coordinates can be expressed as

$$\mathbf{z}_s = (\mathbf{\Lambda}_s^p - \overline{\lambda})^{-1}\mathbf{\Gamma}_s\tau \tag{6}$$

Substitution of Eq. (6) into Eq. (4) gives

$$\begin{bmatrix} \mathbf{\Lambda}_m^p - \overline{\lambda}\mathbf{I} & -\mathbf{\Gamma}_m \\ -\mathbf{\Gamma}_m^T & -\mathbf{\Gamma}_s^T(\mathbf{\Lambda}_s^p - \overline{\lambda}\mathbf{I})^{-1}\mathbf{\Gamma}_s \end{bmatrix} \begin{Bmatrix} \mathbf{z}_m \\ \tau \end{Bmatrix} = \begin{Bmatrix} \mathbf{0} \\ \mathbf{0} \end{Bmatrix} \tag{7}$$

Generally, the lower eigenmodes are usually required by a structure. The eigenvalues $\overline{\lambda}$ are much smaller than $\mathbf{\Lambda}_s^p$ when the size of the master modes is selected rationally. In this regard, Eq. (7) is approximated as:

$$\begin{bmatrix} \mathbf{\Lambda}_m^p - \overline{\lambda}\mathbf{I} & -\mathbf{\Gamma}_m \\ -\mathbf{\Gamma}_m^T & -\mathbf{\Gamma}_s^T(\mathbf{\Lambda}_s^p)^{-1}\mathbf{\Gamma}_s \end{bmatrix} \begin{Bmatrix} \mathbf{z}_m \\ \tau \end{Bmatrix} = \begin{Bmatrix} \mathbf{0} \\ \mathbf{0} \end{Bmatrix} \tag{8}$$

The above eigenequation can be simplified by denoting τ with $\mathbf{z}_m$ from the second line of Eq. (8) and substituting it into the first line as:

$$[(\mathbf{\Lambda}_m^p - \overline{\lambda}\mathbf{I}_m) + \mathbf{\Gamma}_m\zeta^{-1}\mathbf{\Gamma}_m^T]\mathbf{z}_m = \mathbf{0} \tag{9}$$

Consequently, $\overline{\lambda}$ and $\mathbf{z}_m$ are available by solving Eq. (9) with commonly used eigensolver such as Simpson method or Lanczos method [10]. And the eigenvector of the global structure is recovered from the master modes by $\overline{\mathbf{\Phi}} = \mathbf{\Phi}_m^p\mathbf{z}_m$. The size of the simplified eigenequation (Eq. (9)) is equal to the number of the master modes, which is much smaller than the original one (Eq. (2)). It is noted from Eq. (9) that only the master eigensolutions of the independent substructures are used to gain the eigensolutions of the global structure. The contribution of the slave modes is compensated by the first-order residual flexibility $\zeta = \mathbf{\Gamma}_s^T(\mathbf{\Lambda}_s^p)^{-1}\mathbf{\Gamma}_s$, which is calculated by the master modes as:

$$\boldsymbol{\Gamma}_s^T(\boldsymbol{\Lambda}_s^p)^{-1}\boldsymbol{\Gamma}_s = \mathbf{C}\boldsymbol{\Phi}_s^p(\boldsymbol{\Lambda}_s^p)^{-1}[\boldsymbol{\Phi}_s^p]^T\mathbf{C}^T \tag{10}$$

$$\boldsymbol{\Phi}_s^p(\boldsymbol{\Lambda}_s^p)^{-1}[\boldsymbol{\Phi}_s^p]^T = \begin{bmatrix} \left(\mathbf{K}^{(1)}\right)^{-1} - \boldsymbol{\Phi}_m^{(1)}\left(\boldsymbol{\Lambda}_m^{(1)}\right)^{-1}[\boldsymbol{\Phi}_m^{(1)}]^T & & \\ & \ddots & \\ & & \left(\mathbf{K}^{(N_s)}\right)^{-1} - \boldsymbol{\Phi}_m^{(N_s)}\left(\boldsymbol{\Lambda}_m^{(N_s)}\right)^{-1}[\boldsymbol{\Phi}_m^{(N_s)}]^T \end{bmatrix}$$

2.2. Eigensensitivity

In this section, the eigensensitivity of the ith (i=1, 2, …, N) mode with respect to an elemental parameter will be derived. The elemental stiffness parameter α in the Ath substructure is illustrated in the following. Writing Eq. (9) for the ith mode and differentiating it with respect to parameter α gives [11]

$$[(\boldsymbol{\Lambda}_m^p - \overline{\lambda}_i\mathbf{I}_m) + \boldsymbol{\Gamma}_m\zeta^{-1}\boldsymbol{\Gamma}_m^T]\frac{\partial\{\mathbf{z}_i\}}{\partial\alpha} + \frac{\partial[(\boldsymbol{\Lambda}_m^p - \overline{\lambda}_i\mathbf{I}_m) + \boldsymbol{\Gamma}_m\zeta^{-1}\boldsymbol{\Gamma}_m^T]}{\partial\alpha}\{\mathbf{z}_i\} = \{\mathbf{0}\} \tag{11}$$

Premultiplying $\{\mathbf{z}_i\}^T$ on both sides of Eq. (11) gives

$$\{\mathbf{z}_i\}^T[\boldsymbol{\Lambda}_m^p + \boldsymbol{\Gamma}_m\zeta^{-1}\boldsymbol{\Gamma}_m^T - \overline{\lambda}_i\mathbf{I}]\left\{\frac{\partial\mathbf{z}_i}{\partial\alpha}\right\} + \{\mathbf{z}_i\}^T\frac{\partial[\boldsymbol{\Lambda}_m^p + \boldsymbol{\Gamma}_m\zeta^{-1}\boldsymbol{\Gamma}_m^T - \overline{\lambda}_i\mathbf{I}]}{\partial\alpha}\{\mathbf{z}_i\} = 0 \tag{12}$$

Since $[(\boldsymbol{\Lambda}_m^p - \overline{\lambda}\mathbf{I}_m) + \boldsymbol{\Gamma}_m\zeta^{-1}\boldsymbol{\Gamma}_m^T]\mathbf{z}_m = \mathbf{0}$ (Eq. (9)) and $[\boldsymbol{\Lambda}_m^p + \boldsymbol{\Gamma}_m\zeta^{-1}\boldsymbol{\Gamma}_m^T - \overline{\lambda}_i\mathbf{I}]$ are a symmetric matrix, the first item on the left side of Eq. (12) is zero. In consequence, the ith eigenvalue derivative with respect to the designed parameter α is available by [12]

$$\frac{\partial\overline{\lambda}_i}{\partial\alpha} = \{\mathbf{z}_i\}^T\left[\frac{\partial\boldsymbol{\Lambda}_m^p}{\partial\alpha} + \frac{\partial(\boldsymbol{\Gamma}_m\zeta^{-1}\boldsymbol{\Gamma}_m^T)}{\partial\alpha}\right]\{\mathbf{z}_i\} \tag{13}$$

where

$$\frac{\partial(\boldsymbol{\Gamma}_m\zeta^{-1}\boldsymbol{\Gamma}_m^T)}{\partial\alpha} = \frac{\partial\boldsymbol{\Gamma}_m}{\partial\alpha}\zeta^{-1}\boldsymbol{\Gamma}_m^T - \boldsymbol{\Gamma}_m\zeta^{-1}\frac{\partial\zeta}{\partial\alpha}\zeta^{-1}\boldsymbol{\Gamma}_m^T + \boldsymbol{\Gamma}_m\zeta^{-1}\frac{\partial\boldsymbol{\Gamma}_m^T}{\partial\alpha} \tag{14}$$

$\frac{\partial\boldsymbol{\Lambda}_m^p}{\partial\alpha}$ and $\frac{\partial\boldsymbol{\Gamma}_m}{\partial\alpha}$ are the eigenvalue and eigenvector derivatives of the master modes of the independent substructures, respectively. $\frac{\partial\zeta}{\partial\alpha} = \frac{\partial\left(\boldsymbol{\Gamma}_s^T(\boldsymbol{\Lambda}_s^p)^{-1}\boldsymbol{\Gamma}_s\right)}{\partial\alpha}$ is the derivative of the residual flexibility of the substructures. Considering that the substructures are taken as independent structures, these derivative matrices are calculated within the Ath substructure solely, while the corresponding derivative matrices in other substructures are zero matrices, i.e.,

$$\frac{\partial\boldsymbol{\Lambda}_m^p}{\partial\alpha} = \begin{bmatrix} \mathbf{0} & \mathbf{0} & \mathbf{0} \\ \mathbf{0} & \frac{\partial\boldsymbol{\Lambda}_m^{(A)}}{\partial\alpha} & \mathbf{0} \\ \mathbf{0} & \mathbf{0} & \mathbf{0} \end{bmatrix}, \frac{\partial\boldsymbol{\Gamma}_m^T}{\partial\alpha} = \mathbf{C}\frac{\partial\boldsymbol{\Phi}_m^p}{\partial\alpha} = \mathbf{C}\begin{bmatrix} \mathbf{0} & \mathbf{0} & \mathbf{0} \\ \mathbf{0} & \frac{\partial\boldsymbol{\Phi}_m^{(A)}}{\partial\alpha} & \mathbf{0} \\ \mathbf{0} & \mathbf{0} & \mathbf{0} \end{bmatrix}$$
$$\frac{\partial\zeta}{\partial\alpha} = \frac{\partial\left[\left(\boldsymbol{\Gamma}_s^T(\boldsymbol{\Lambda}_s^p)^{-1}\boldsymbol{\Gamma}_s\right)^{-1}\right]}{\partial\alpha} = \mathbf{C}\begin{bmatrix} \mathbf{0} & \mathbf{0} & \mathbf{0} \\ \mathbf{0} & \frac{\partial\left(\left(\mathbf{K}^{(A)}\right)^{-1} - \boldsymbol{\Phi}_m^{(A)}\left(\boldsymbol{\Lambda}_m^{(A)}\right)^{-1}[\boldsymbol{\Phi}_m^{(A)}]^T\right)}{\partial\alpha} & \mathbf{0} \\ \mathbf{0} & \mathbf{0} & \mathbf{0} \end{bmatrix}\mathbf{C}^T \tag{15}$$

$\frac{\partial \Lambda_m^{(A)}}{\partial \alpha}$ and $\frac{\partial \Phi_m^{(A)}}{\partial \alpha}$ can be calculated rapidly by treating the Ath substructure as an independent structure with Nelson's method [12, 13].

The ith eigenvector of the global structure is recovered by the master modes as

$$\overline{\mathbf{\Phi}}_i = \mathbf{\Phi}_m^p\{\mathbf{z}_i\} \tag{16}$$

Eq. (16) is differentiated with respect to the structural parameter α as

$$\frac{\partial \overline{\mathbf{\Phi}}_i}{\partial \alpha} = \frac{\partial \mathbf{\Phi}_m^p}{\partial \alpha}\{\mathbf{z}_i\} + \mathbf{\Phi}_m^p\left\{\frac{\partial \mathbf{z}_i}{\partial \alpha}\right\} \tag{17}$$

where $\mathbf{\Phi}_m^p$ and $\frac{\partial \mathbf{\Phi}_m^p}{\partial \alpha}$ are the master eigenvectors and their derivatives of the Ath substructure, respectively. $\{\mathbf{z}_i\}$ is the eigenvector calculated from Eq. (9). Only $\left\{\frac{\partial \mathbf{z}_i}{\partial \alpha}\right\}$ is required to calculate the eigenvector derivative of the ith mode in Eq. (17).

$\left\{\frac{\partial \mathbf{z}_i}{\partial \alpha}\right\}$ is rewritten by the sum of a particular part and a general part as

$$\left\{\frac{\partial \mathbf{z}_i}{\partial \alpha}\right\} = \{v_i\} + c_i\{\mathbf{z}_i\} \tag{18}$$

where c_i is a participation factor and $\{v_i\}$ is a residual vector. Substituting Eq. (18) into Eq. (11) leads to

$$[\mathbf{\Lambda}_m^p + \mathbf{\Gamma}_m\zeta^{-1}\mathbf{\Gamma}_m^T - \overline{\lambda}_i\mathbf{I}](\{v_i\} + c_i\{\mathbf{z}_i\}) = -\frac{\partial[\mathbf{\Lambda}_m^p + \mathbf{\Gamma}_m\zeta^{-1}\mathbf{\Gamma}_m^T - \overline{\lambda}_i\mathbf{I}]}{\partial r}\{\mathbf{z}_i\} \tag{19}$$

Given that $[\mathbf{\Lambda}_m^p + \mathbf{\Gamma}_m\zeta^{-1}\mathbf{\Gamma}_m^T - \overline{\lambda}_i\mathbf{I}]\{\mathbf{z}_i\} = \{\mathbf{0}\}$, Eq. (19) can be simplified into

$$\mathbf{\Psi}\{v_i\} = \{Y_i\} \tag{20}$$

where

$$\mathbf{\Psi} = [\mathbf{\Lambda}_m^p + \mathbf{\Gamma}_m\zeta^{-1}\mathbf{\Gamma}_m^T - \overline{\lambda}_i\mathbf{I}],\ \{Y_i\} = -\frac{\partial[\mathbf{\Lambda}_m^p + \mathbf{\Gamma}_m\zeta^{-1}\mathbf{\Gamma}_m^T - \overline{\lambda}_i\mathbf{I}]}{\partial \alpha}\{\mathbf{z}_i\} \tag{21}$$

In consequence, $\mathbf{\Psi}$ and $\{Y_i\}$ can be calculated from Eq. (21) since all of their items have been available in the calculation of the eigenvalue derivatives proposed in the former section.

If no repeated roots exist in Eq. (20), $\mathbf{\Psi}$ takes the size of $m^p \times m^p$ with the rank of (m^p-1). To solve this rank-deficient equation (Eq. (20)), the kth item (corresponds to the maximum entry in $\{\mathbf{z}_i\}$) in $\{v_i\}$ is assumed to be zero, and the corresponding row and column in $\mathbf{\Psi}$ and corresponding item in $\{Y_i\}$ are assumed to be zeros as well [14]. The full rank equation is formed as

$$\begin{bmatrix} \mathbf{\Psi}_{11} & \mathbf{0} & \mathbf{\Psi}_{13} \\ \mathbf{0} & 1 & \mathbf{0} \\ \mathbf{\Psi}_{31} & \mathbf{0} & \mathbf{\Psi}_{33} \end{bmatrix} \begin{Bmatrix} \nu_{i1} \\ \nu_{ik} \\ \nu_{i3} \end{Bmatrix} = \begin{Bmatrix} Y_{i1} \\ 0 \\ Y_{i3} \end{Bmatrix} \tag{22}$$

In consequence, the vector $\{\nu_i\}$ is solved from Eq. (22).

The eigenvectors $\{\mathbf{z}_i\}$ satisfy the orthogonal condition of

$$\{\mathbf{z}_i\}^T\{\mathbf{z}_i\} = 1 \tag{23}$$

Equation (23) is differentiated with respect to α as

$$\frac{\partial\{\mathbf{z}_i\}^T}{\partial\alpha}\{\mathbf{z}_i\} + \{\mathbf{z}_i\}^T\frac{\partial\{\mathbf{z}_i\}}{\partial\alpha} = 0 \tag{24}$$

Substitution of Eq. (18) into Eq. (24) gives

$$(\{\nu_i\}^T + c_i\{\mathbf{z}_i\}^T)\{\mathbf{z}_i\} + \{\mathbf{z}_i\}^T(\{\nu_i\} + c_i\{\mathbf{z}_i\}) = 0 \tag{25}$$

The participation factor c_i is thus obtained as

$$c_i = -\frac{1}{2}(\{\nu_i\}^T\{\mathbf{z}_i\} + \{\mathbf{z}_i\}^T\{\nu_i\}) \tag{26}$$

Finally, the first-order derivative of $\{\mathbf{z}_i\}$ with respect to α is calculated by

$$\left\{\frac{\partial\mathbf{z}_i}{\partial\alpha}\right\} = \{\nu_i\} - \frac{1}{2}(\{\nu_i\}^T\{\mathbf{z}_i\} + \{\mathbf{z}_i\}^T\{\nu_i\})\{\mathbf{z}_i\} \tag{27}$$

It is noted from Eq. (17) that the eigenvector derivatives of the global structure are calculated from $\mathbf{\Phi}_m^p$ and $\frac{\partial\mathbf{\Phi}_m^p}{\partial\alpha}$. $\left\{\frac{\partial\mathbf{z}_i}{\partial\alpha}\right\}$ and $\{\mathbf{z}\}$ are treated as the weights and are computed from the small-size eigenequation (Eq.(9)) rapidly. Only the derivative matrices of the master modes in the Ath substructure are needed to recover the eigensensitivity of the global structure. As the size of the independent substructures is much smaller than that of the global structure, the proposed substructuring method can significantly improve the computational efficiency.

2.3. Substructure-based updating method

Based on the eigensolutions and eigensensitivities calculated with the forward substructuring method, the substructure-based model updating is described in **Figure 1** with an iterative process. In each iteration, the eigensolutions are calculated from the modified substructures with the above substructuring method and are then compared with the experimental modal data (frequencies and mode shapes) to construct the objective function. The substructure-based eigensensitivities with respect to a specific parameter are calculated from the substructure containing the concerned parameter, to indicate the searching direction in each optimal

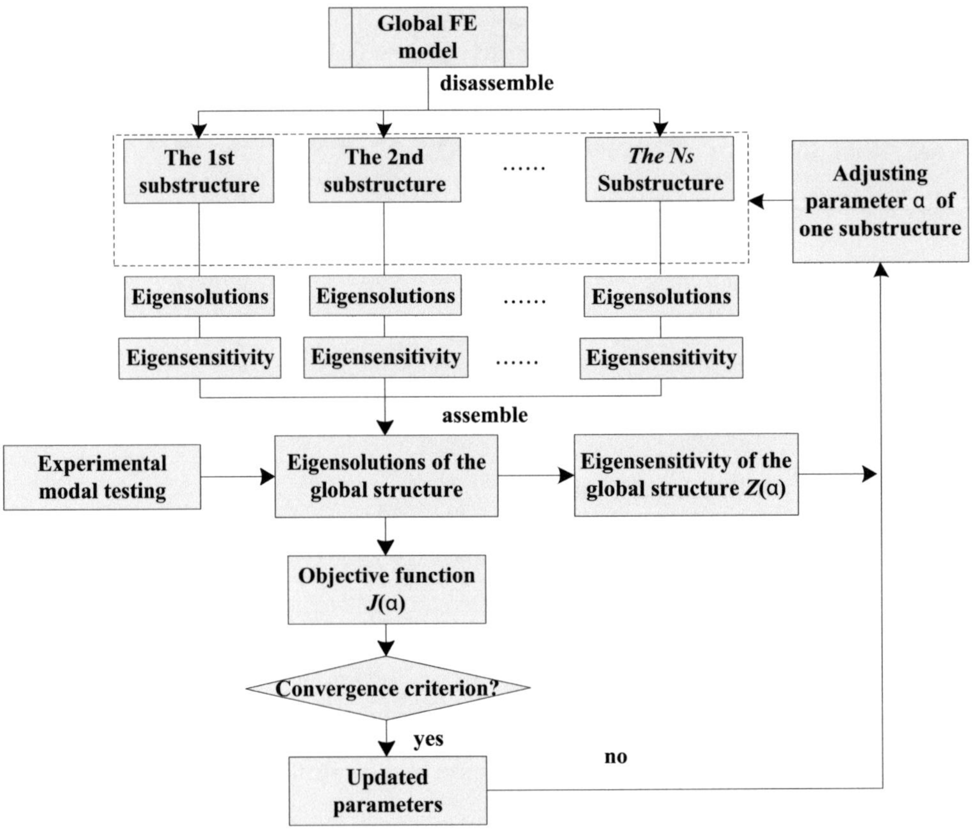

Figure 1. The model updating of forward substructuring method.

step. The objective function is minimized by adjusting the elemental parameters α iteratively according to the eigensensitivity matrices.

The objective function formed by the modal frequency and the mode shape is written as [14]

$$J(\alpha) = \sum_i W_{\lambda i}^2 \left[\lambda_i(\{\alpha\})^{FE} - \lambda_i{}^E\right]^2 + \sum_i W_{\phi i}^2 \sum_j \left[\phi_{ji}(\{\alpha\})^{FE} - \phi_{ji}{}^E\right]^2 \tag{28}$$

where $\lambda_i{}^E$ and $\phi_{ji}{}^E$ represent the experimental frequencies and mode shapes, respectively. $\lambda_i{}^{FE}$ and $\phi_{ji}{}^{FE}$ are the frequencies and mode shapes gained from the analytical FE model with the substructuring method (Eq.(8)) proposed above. $W_{\lambda i}$ and $W_{\phi i}$ are the weighting matrix of frequencies and mode shapes. The objective function is minimized by adjusting the elemental parameters α in an optimal manner.

The eigensensitivity is computed with the first derivative of a structural response with respect to a physical parameter as [2]

$$[S_\lambda(\alpha)] = \frac{\partial\lambda(\alpha)}{\partial\alpha}, [S_\phi(\alpha)] = \frac{\partial\phi(\alpha)}{\partial\alpha} \tag{29}$$

In this chapter, the eigensensitivity matrices are available with the forward substructuring method. They are computed solely from the derivative matrices of the substructure containing the concerned element, while the corresponding derivative matrices of all other substructures are zeros. As the calculation of eigensensitivity usually consumes most of the computation time when numerous elemental parameters are updated in practical model updating process, the forward substructuring method can significantly improve the computational efficiency of the model-updating process.

3. Inverse substructuring method

3.1. The extraction of substructural flexibility

In the inverse substructuring method, the global flexibility matrix estimated from the experimental modal data is disassembled into substructural flexibility matrices. Afterwards, the analytical FE models of the substructures are updated independently and parallelly to reproduce the extracted substructural flexibility matrices. As before, the global structure with N DOFs is divided into Ns independent substructures with the jth ($j = 1, 2,\ldots, Ns$) substructure $n^{(j)}$ DOFs. Treated as independent substructures, the substructural displacements, forces, stiffness, flexibility, and rigid body modes matrices are written in the primitive form as

$$\begin{aligned}&\{x^p\} = \left\{x^{(1)}\cdots x^{(j)}\cdots x^{(N_s)}\right\}^T, \{f^p\} = \left\{f^{(1)}\cdots f^{(j)}\cdots f^{(N_s)}\right\}^T\\&\mathbf{K}^p = Diag\left[\mathbf{K}^{(1)}\cdots\mathbf{K}^{(j)}\cdots\mathbf{K}^{(N_s)}\right], \mathbf{F}^p = Diag\left[\mathbf{F}^{(1)}\cdots\mathbf{F}^{(j)}\cdots\mathbf{F}^{(N_s)}\right], \mathbf{R}^p = Diag\left[\mathbf{R}^{(1)}\cdots\mathbf{R}^{(j)}\cdots\mathbf{R}^{(N_s)}\right]\end{aligned} \tag{30}$$

where $\mathbf{K}^{(j)}$, $\mathbf{F}^{(j)}$, $x^{(j)}$, $f^{(j)}$, and $\mathbf{R}^{(j)}$, respectively, represent the stiffness, flexibility, nodal displacements, external forces, and rigid body modes of the jth substructure. It is noted that the rigid body modes $\mathbf{R}$ is related to free-constraint substructures. R is a zero matrix if the jth substructure is constrained after partition. Otherwise, $\mathbf{R}$ is determined by the nodal location. For example, a two-dimensional structure with n nodes has three rigid body modes, i.e., the x translation ($\mathbf{R}_x = 1$, $\mathbf{R}_y = 0$), the y translation ($\mathbf{R}_x = 0$, $\mathbf{R}_y = 1$) and the z rotation ($\mathbf{R}_x = -y$, $\mathbf{R}_y = x$), $\mathbf{R}$ takes the form of

$$\mathbf{R}^T = \begin{bmatrix} 1 & 0 & 0 & 1 & \cdots & 0 & 0 \\ 0 & 1 & 0 & 0 & \cdots & 1 & 0 \\ -y_1 & x_1 & 1 & -y_2 & \cdots & x_n & 1 \end{bmatrix} \tag{31}$$

The primitive forms of the substructural displacements and forces are associated with the global counterparts as [15]

$$\{x^p\} = \mathbf{L}^p\{x_g\}, [\mathbf{L}^p]^T\{f^p\} = \{f_g\} \tag{32}$$

where $\{x_g\}$ and $\{f_g\}$ are the nodal displacement and external force vector of the global structure. $\mathbf{L}^p$ is a Boolean matrix composed of 1 and 0 values to relate the DOFs of the substructures and the global structure [5]. Most of the values in $\mathbf{L}^p$ are zeros. $\mathbf{L}^p_{ij} = 1$ means that the jth DOF of the global structure corresponds to the ith DOF in the partitioned substructures. The displacement of an independent substructure is constituted by its deformational motions and rigid body motion

$$\{x^p\} = \mathbf{F}^p\{f^p\} + \mathbf{R}^p\{\beta^p\} \tag{33}$$

where β is the participation factor of rigid body modes. As an independent structure, a substructure is excited by the external force and the internal interface force from the adjacent substructures as

$$\{f^p\} = ([\mathbf{L}^p]^T)^+\{f_g\} + \mathbf{C}\{\tau\} = \{\tilde{f}_g\} + \mathbf{C}\{\tau\} \tag{34}$$

where $\{\tilde{f}_g\} = ([\mathbf{L}^p]^T)^+\{f_g\} = \tilde{\mathbf{L}}^p\{f_g\}$, $\tilde{\mathbf{L}}^p = ([\mathbf{L}^p]^T)^+$is the generalized inverse of $[\mathbf{L}^p]^T$. Similar to the forward substructuring method, $\{\tau\}$ denotes the internal interface forces from the adjacent substructures, and matrix $\mathbf{C}$ implicitly defines the connections between the adjacent substructures. Substitution of Eq. (34) into Eq. (33) gives

$$\{x^p\} = \mathbf{F}^p(\{\tilde{f}_g\} + \mathbf{C}\{\tau\}) + \mathbf{R}^p\{\beta^p\} \tag{35}$$

Substitution of Eq. (35) into the left equation of Eq. (32) gives

$$\{x_g\} = [\mathbf{L}^p]^+\{x^p\} = [\tilde{\mathbf{L}}^p]^T\mathbf{F}^p(\{\tilde{f}_g\} + \mathbf{C}\{\tau\}) + [\tilde{\mathbf{L}}^p]^T\mathbf{R}^p\{\beta^p\} \tag{36}$$

Since the global displacement is associated with the global force by $\{x_g\} = \mathbf{F}_g\{f_g\}$ [15], the global flexibility can also be expressed as

$$\{x_g\} = [\mathbf{L}^p]^+\{x^p\} = [\tilde{\mathbf{L}}^p]^T\mathbf{F}^p(\{\tilde{f}_g\} + \mathbf{C}\{\tau\}) + [\tilde{\mathbf{L}}^p]^T\mathbf{R}^p\{\beta^p\} = \mathbf{F}_g\{f_g\} \tag{37}$$

Equation (37) means that the primitive substructural flexibility matrix $\mathbf{F}^p$ can be calculated from the global flexibility matrix $\mathbf{F}_g$ once the two variables $\{\tau\}$ and $\{\beta^p\}$ are given. $\{\tau\}$ and $\{\beta^p\}$ are gained according the force and displacement compatibility condition with the following procedures:

1. The primitive substructural rigid body modes and forces satisfy the force equilibrium compatibility as [16, 17]>

$$[\mathbf{R}^p]^T\{f^p\} = \{\mathbf{0}\} \tag{38}$$

2. From the physical point of view, matrix $\mathbf{C}$ constraints the displacement compatibility as

$$\mathbf{C}^T\{x^p\} = \{\mathbf{0}\} \tag{39}$$

Substituting Eqs. (33) and (34) into Eq. (39) leads to

$$\mathbf{C}^T\{\mathbf{F}^p(\{\tilde{f}_g\}+\mathbf{C}\{\tau\})+\mathbf{R}^p\{\beta^p\}\}=\{\mathbf{0}\} \tag{40}$$

Therefore, $\{\tau\}$ is expressed as

$$\{\tau\}=-\mathbf{F}_C^{-1}(\mathbf{C}^T\mathbf{F}^p\{\tilde{f}_g\}+\mathbf{R}_C\{\beta^p\}) \tag{41}$$

where $\mathbf{F}_C=\mathbf{C}^T\mathbf{F}^p\mathbf{C}$ and $\mathbf{R}_C=\mathbf{C}^T\mathbf{R}^p$.

The combination of Eq. (34) and Eq. (41) gives

$$[\mathbf{R}^p]^T\left(\{\tilde{f}_g\}-\mathbf{C}\mathbf{F}_C^{-1}(\mathbf{C}^T\mathbf{F}^p\{\tilde{f}_g\}+\mathbf{R}_C\{\beta^p\})\right)=\{\mathbf{0}\} \tag{42}$$

$\{\beta^p\}$ is therefore solved as

$$\{\beta^p\}=\mathbf{K}_R^{-1}([\mathbf{R}^p]^T-\mathbf{R}_C^T\mathbf{F}_C^{-1}\mathbf{C}^T\mathbf{F}^p)\{\tilde{f}_g\} \tag{43}$$

where $\mathbf{K}_R=\mathbf{R}_C^T\mathbf{F}_C^{-1}\mathbf{R}_C$. In consequence, $\{\tau\}$ is therefore solved from Eq. (41) as

$$\{\tau\}=-\mathbf{F}_C^{-1}\mathbf{C}^T\mathbf{F}^p\{\tilde{f}_g\}+\mathbf{F}_C^{-1}\mathbf{C}^T\mathbf{R}^p\mathbf{K}_R^{-1}([\mathbf{R}^p]^T\mathbf{K}_C\mathbf{F}^p-[\mathbf{R}^p]^T)\{\tilde{f}_g\} \tag{44}$$

where $\mathbf{K}_C=\mathbf{C}\mathbf{F}_C^{-1}\mathbf{C}^T$. Once $\{\tau\}$ and $\{\beta^p\}$ are solved, Eq. (36) can be expressed as

$$\{x_g\}=[\tilde{\mathbf{L}}^p]^T(\mathbf{F}^p-\mathbf{F}^p\mathbf{H}\mathbf{F}^p-\mathbf{F}^p\mathbf{K}_C\mathbf{F}_R-\mathbf{F}_R^T\mathbf{K}_C^T\mathbf{F}^p+\mathbf{F}_R)\tilde{\mathbf{L}}^p\{f_g\} \tag{45}$$

where

$$\mathbf{F}_R=\mathbf{R}^p([\mathbf{R}^p]^T\mathbf{K}_C\mathbf{R}^p)^{-1}[\mathbf{R}^p]^T,\mathbf{H}=\mathbf{K}_C-\mathbf{K}_C\mathbf{F}_R\mathbf{K}_C$$

In consequence, the global flexibility matrix can be expressed by the substructural flexibility matrix:

$$\mathbf{L}^p\mathbf{F}_g[\mathbf{L}^p]^T=\mathbf{F}^p-\mathbf{F}^p\mathbf{K}_C\mathbf{F}_R-\mathbf{F}_R\mathbf{K}_C\mathbf{F}^p-\mathbf{F}^p\mathbf{H}\mathbf{F}^p+\mathbf{F}_R \tag{46}$$

Based on Eq. (46), the substructural flexibility matrix $\mathbf{F}^p$ is extracted from the global flexibility $\mathbf{F}_g$ with an iterative scheme:

1. $\mathbf{F}^p$ is initiated from the diagonal subblocks of the global flexibility as

$$[\mathbf{F}^p]^{[0]}=\mathbf{L}^p\begin{bmatrix}\mathbf{F}_{\left(1:N^{(1)},1:N^{(1)}\right)} & & & & \\ & \ddots & & & \\ & & \mathbf{F}_{\left(\sum_{i=1}^{j-1}N^{(i)}+1:\sum_{i=1}^{j}N^{(i)},\sum_{i=1}^{j-1}N^{(i)}+1:\sum_{i=1}^{j}N^{(i)}\right)} & & \\ & & & \ddots & \\ & & & & \mathbf{F}_{\left(\sum_{i=1}^{N_s-1}N^{(i)}+1:\sum_{i=1}^{N_s}N^{(i)},\sum_{i=1}^{N_s-1}N^{(i)}+1:\sum_{i=1}^{N_s}N^{(i)}\right)}\end{bmatrix}[\mathbf{L}^p]^T \tag{47}$$

2. In the kth ($k = 1, 2, \ldots$) iteration, the substructural flexibility matrix is calculated according to Eq. (46)

$$[\mathbf{F}^p]_0^{[k]} = \tilde{\mathbf{F}}_g + [\mathbf{F}^p]^{[k-1]}\mathbf{H}^{[k-1]}[\mathbf{F}^p]^{[k-1]} + [\mathbf{F}^p]^{[k-1]}\mathbf{K}_C^{[k-1]}\mathbf{F}_R^{[k-1]} + \mathbf{F}_R^{[k-1]}\mathbf{K}_C^{[k-1]}[\mathbf{F}^p]^{[k-1]} - \mathbf{F}_R^{[k-1]} \quad (48)$$

The diagonal subblocks of $[\mathbf{F}_0^p]^{[k]}$ are reused in the next iteration

$$[\mathbf{F}^p]^{[k]} = \begin{bmatrix} [\mathbf{F}^p]_0^{[k]}\left(_{1:N^{(1)},1:N^{(1)}}\right) & & & & \\ & \ddots & & & \\ & & [\mathbf{F}_0^p]_0^{[k]}\left(\sum_{i=1}^{j-1}N^{(i)}{}_{+1:}\sum_{i=1}^{j}N^{(i)},\sum_{i=1}^{j-1}N^{(i)}{}_{+1:}\sum_{i=1}^{j}N^{(i)}\right) & & \\ & & & \ddots & \\ & & & & [\mathbf{F}_0^p]_0^{[k]}\left(\sum_{i=1}^{N_s-1}N^{(i)}{}_{+1:}\sum_{i=1}^{N_s}N^{(i)},\sum_{i=1}^{N_s-1}N^{(i)}{}_{+1:}\sum_{i=1}^{N_s}N^{(i)}\right) \end{bmatrix} \quad (49)$$

3. Step 2 stops when the substructural flexibility matrices from two consecutive iterations drop below a predefined tolerance [16]

$$e = \frac{norm([\mathbf{F}^p]^{[k]} - [\mathbf{F}^p]^{[k-1]})}{norm([\mathbf{F}^p]^{[k]})} \leq Tol \quad (50)$$

The substructural flexibility matrices $\mathbf{F}^{(j)}$ are thereby gained by the diagonal subblocks of $[\mathbf{F}^p]^{[k]}$.

3.2. The projection matrix to extract free-free flexibility for model updating

In the substructuring methods, the global structure is divided properly into several independent free or constrained substructures. Most of the substructures are free-free without constraints after partition. Here the jth substructure is free-free as an illustration. The substructural flexibility matrix $\overline{\mathbf{F}}^{(j)}$ from $\mathbf{F}^p$ is constituted by both the rigid body modes and deformational modes. Hereinafter, superscript "j" is omitted to derive the free-free substructural flexibility for brevity. For the jth substructure, the substructural flexibility matrix, contributed by the rigid body motions and deformational motions, is expressed as

$$\overline{\mathbf{F}} = \mathbf{F} + \gamma\mathbf{R}\mathbf{R}^T \quad (51)$$

$\overline{\mathbf{F}}$ is defined as the generalized substructural flexibility. Accordingly, the generalized substructural stiffness matrix, including the contribution made by the rigid body motions and deformational motions is written as

$$\overline{\mathbf{K}} = \mathbf{K} + \eta \mathbf{R}\mathbf{R}^T \tag{52}$$

where $\overline{\mathbf{K}}$ is defined as the generalized substructural stiffness matrix. The free-free stiffness and flexibility matrices (**K** and **F**) are contributed by the deformational modes solely. The participation factors γ and η of rigid body modes are difficult to determine, which makes the generalized flexibility unable to be applied to model updating or damage identification. It is necessary to extract the free-free substructural flexibility contributed by the deformational modes solely. The free-free flexibility shows the real properties of a substructure and can be applied to model updating and damage identification.

To remove the rigid body components in the generalized substructural stiffness, flexibility, and displacements, a projection matrix **P** is formed as [17]

$$\mathbf{P} = \mathbf{I} - \mathbf{R}(\mathbf{R}^T\mathbf{R})^{-1}\mathbf{R}^T \tag{53}$$

The projection matrix **P** has the properties of

$$\mathbf{P}^2 = \mathbf{P}, \mathbf{P}\mathbf{R} = \mathbf{R}^T\mathbf{P} = \mathbf{0} \tag{54}$$

P can filter out the rigid body motions, while the free-free stiffness and flexibility matrices contributed by the deformational modes remain unchanged

$$\begin{aligned}
&\mathbf{F}\mathbf{P} = \mathbf{F}, \mathbf{P}\mathbf{F} = \mathbf{F}, \mathbf{P}\mathbf{F}\mathbf{P} = \mathbf{F} \\
&\overline{\mathbf{F}}\mathbf{P} = \mathbf{F}, \mathbf{P}\overline{\mathbf{F}} = \mathbf{F}, \mathbf{P}\overline{\mathbf{F}}\mathbf{P} = \mathbf{F} \\
&\mathbf{K}\mathbf{P} = \mathbf{K}, \mathbf{P}\mathbf{K} = \mathbf{K}, \mathbf{P}^T\mathbf{K}\mathbf{P} = \mathbf{K} \\
&\overline{\mathbf{K}}\mathbf{P} = \mathbf{K}, \mathbf{P}^T\overline{\mathbf{K}} = \mathbf{K}, \mathbf{P}^T\overline{\mathbf{K}}\mathbf{P} = \mathbf{K}
\end{aligned} \tag{55}$$

On the other hand, the free-free stiffness and flexibility of a substructural analytical model are singular, whereas the generalized stiffness and flexibility are full-rank. The free-free stiffness and flexibility can be calculated from the inverse of the generalized stiffness and flexibility matrices as

$$\mathbf{F} = \mathbf{P}(\mathbf{K} + \eta\mathbf{R}\mathbf{R}^T)^{-1}\mathbf{P} \tag{56}$$

$$\mathbf{K} = \mathbf{P}(\mathbf{F} + \gamma\mathbf{R}\mathbf{R}^T)^{-1}\mathbf{P} \tag{57}$$

If the projection matrix **P** is known, the free-free substructural flexibility **F** is calculated from Eq. (56) or by removing all the rigid body components in the extracted substructural flexibility matrix (Eq. (55)). In substructure-based model updating, the elemental parameters of the analytical FE model are iteratively adjusted to minimize the discrepancy between the analytical substructural flexibility and that extracted from global data [18].

Generally, the stiffness or flexibility matrices are difficult to be measured on the full DOFs, and the partial stiffness and flexibility at the measured DOFs are probably utilized for a substructure. Divide the full-DOF model into the measured part and the unmeasured part, the stiffness matrix is rewritten in block form as

$$\mathbf{K} = \begin{bmatrix} \mathbf{K}_{aa} & \mathbf{K}_{ab} \\ \mathbf{K}_{ba} & \mathbf{K}_{bb} \end{bmatrix} \tag{58}$$

where subscript "*a*" represents the measured DOFs, and subscript "*b*" represents the unmeasured DOFs. The condensed stiffness matrix by the Guyan static condensation is [19–21]

$$\mathbf{K}_G = \mathbf{K}_{aa} - \mathbf{K}_{ab}\mathbf{K}_{bb}^{-1}\mathbf{K}_{ba} \tag{59}$$

The substructural flexibility is written in block form according to the measured and unmeasured parts as [22]

$$\mathbf{F} = \begin{bmatrix} \mathbf{F}_{aa} & \mathbf{F}_{ab} \\ \mathbf{F}_{ba} & \mathbf{F}_{bb} \end{bmatrix}, \overline{\mathbf{F}} = \begin{bmatrix} \overline{\mathbf{F}}_{aa} & \overline{\mathbf{F}}_{ab} \\ \overline{\mathbf{F}}_{ba} & \overline{\mathbf{F}}_{bb} \end{bmatrix} \tag{60}$$

In this case, the projection matrix of the reduced model $\mathbf{P}_D$ is formed as

$$\mathbf{P}_D = \mathbf{I} - \mathbf{R}_a(\mathbf{R}_a^T\mathbf{R}_a)^{-1}\mathbf{R}_a^T \tag{61}$$

which has the properties of

$$\mathbf{P}_D^2 = \mathbf{P}_D \tag{62}$$

$$\mathbf{P}_D\mathbf{R}_a = \mathbf{R}_a^T\mathbf{P}_D = \mathbf{0} \tag{63}$$

The rigid body modes $\mathbf{R}_a$ are gained by rewriting the rows in Eq. (58) corresponding to the measured DOFs.

The projection matrix $\mathbf{P}_D$ removes the rigid body components in the partial substructural flexibility matrix and leaves the free-free substructural flexibility by

$$\mathbf{F}_{aa}\mathbf{P}_D = \mathbf{P}_D\mathbf{F}_{aa} = \mathbf{P}_D\mathbf{F}_{aa}\mathbf{P}_D = \mathbf{F}_{aa} \tag{64}$$

$$\overline{\mathbf{F}}_{aa}\mathbf{P}_D = \mathbf{P}_D\overline{\mathbf{F}}_{aa} = \mathbf{P}_D\overline{\mathbf{F}}_{aa}\mathbf{P}_D = \mathbf{F}_{aa} \tag{65}$$

In addition, the projection matrix can be used to form the dual inverse of substructural stiffness and flexibility like

$$\mathbf{F}_{aa} = \mathbf{P}_D\left(\mathbf{K}_G + \mathbf{R}_a(\mathbf{R}_a^T\mathbf{R}_a)^{-1}\mathbf{R}_a^T\right)^{-1}\mathbf{P}_D \tag{66}$$

$$\mathbf{K}_G = \mathbf{P}_D\left(\mathbf{F}_{aa} + \mathbf{R}_a(\mathbf{R}_a^T\mathbf{R}_a)^{-1}\mathbf{R}_a^T\right)^{-1}\mathbf{P}_D \tag{67}$$

In substructure-based model updating, the elemental parameters in the substructural model are iteratively adjusted to minimize the discrepancy between the substructural flexibility and that extracted from global modal data [18]. For a free-free substructure, the flexibility extracted from global modal data is contaminated by the rigid body motions, and the stiffness matrix of substructural analytical FE model is singular. The projection matrix is utilized to extract the free-free flexibility for model updating. On the one hand, the projection matrix removes the rigid body components in the generalized substructural flexibility from experimental data and leaves the free-free substructural flexibility according to

$$\mathbf{F}_{aa}^{E} = \mathbf{P}_D \overline{\mathbf{F}}_{aa}^{E} \mathbf{P}_D \tag{68}$$

On the other hand, the free-free flexibility matrix of the substructural FE model is iteratively computed from the singular stiffness matrix according to

$$\mathbf{F}_{aa}^{FE} = \mathbf{P}_D \left(\mathbf{K}_G + \mathbf{R}_a (\mathbf{R}_a^T \mathbf{R}_a)^{-1} \mathbf{R}_a^T \right)^{-1} \mathbf{P}_D \tag{69}$$

3.3. Substructure-based model updating

The substructure-based model updating process is listed in **Figure 2**. Identically, the *j*th substructure, which is free-free after partition, is employed to illustrate the substructure-based model updating in the following:

1. The experimental flexibility $\mathbf{F}_g^E$ is estimated by modal data of the global structure.

2. The generalized substructural flexibility matrix $\left(\overline{\mathbf{F}}^{(j)}\right)^E$ is extracted from the global flexibility matrix $\mathbf{F}_g^E$ by the proposed substructuring method in Section 3.1.

3. The rigid body modes $\mathbf{R}_a^{(j)}$ are constructed according to the nodal location of the *j*th substructure (Eq. (31)), and the projection matrix $\mathbf{P}^{(j)}$ is formed according to the proposed method in Section 3.2.

   The free-free substructural flexibility is extracted by the projection matrix as $\left(\mathbf{F}^{(j)}\right)^E = [\mathbf{P}^{(j)}]^T \left(\overline{\mathbf{F}}^{(j)}\right)^E \mathbf{P}^{(j)}$.

4. The FE model of the *j*th substructure is constructed without constraints. The FE model of the *j*th substructure is treated as an independent structure to be updated: In each iteration, the free-free substructural flexibility matrix $\left(\mathbf{F}^{(j)}\right)^{FE}$ at the measured DOFs and its sensitivity with respect to α $\partial\left(\mathbf{F}^{(j)}\right)^{FE}/\partial\alpha$ are computed [21]. The elemental parameters in the *j*th substructure are adjusted according to the sensitivity ($J(\alpha)$) of the flexibility with respect to elemental parameters, to minimize the objective function $\Delta\mathbf{F}(\alpha)$ through the Trust Region Newton method [2, 3, 18].

In the proposed substructuring method, the substructural flexibility matrices in primitive matrix $\mathbf{F}^p$ are independent. And only one substructure instead of the whole global structure at a time is updated in each iteration. The size of system matrices and updating parameters are sharply reduced, which improves the computational efficiency of model updating significantly.

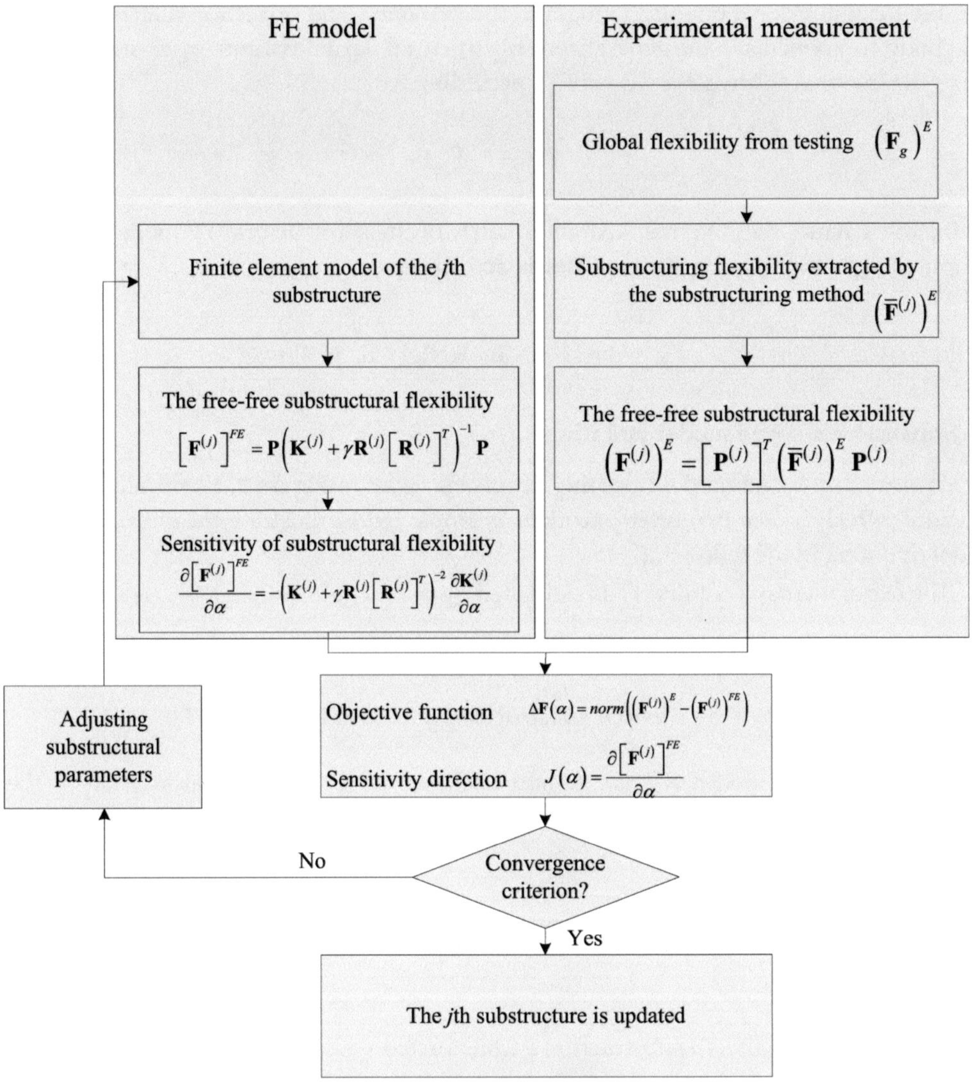

Figure 2. The model updating of inverse substructuring method.

4. Laboratory frame structure

Here a laboratory-tested steel frame structure is employed to investigate the effectiveness of the forward and inverse substructuring methods in model updating and damage identification. The cross section of the beams is 50.0×8.8 mm^2 and the cross section of the columns is 50.0×4.4 mm^2, with the dimensions shown in **Figures 3(a)** and **(b)**. The mass density of the structural material is 7.67×10^3 kg/m^3. The FE model of the frame is composed of 44 nodes and 45 elements, with each element 100 mm in length as **Figure 3(c)**. In experiment, the accelerometers are placed at the nodes to measure the translational vibration of the frame [23]. The sampling

http://dx.doi.org/10.5772/67890

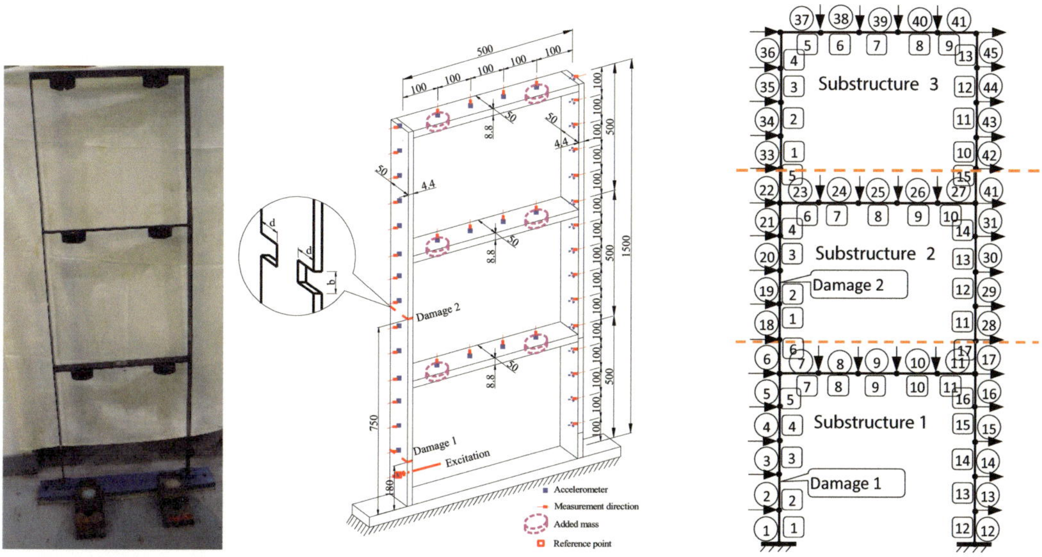

(a) Experimental specimen (b) Configurations (c) Analytical model

Figure 3. Laboratory-tested frame structure. (a) Experimental specimen. (b) Configurations. (c) Analytical model.

frequency was set to 2000 Hz. The specimen was excited with the instrumented hammer at the reference point indicated in **Figure 3(a)**.

The FE model is first updated in the undamaged state, and the refined model is subsequently used for damage identification. In the undamaged state, the Young's modules of all 45 elements are updated, with their initial values set to 2×10^{11} Pa. The global structure is partitioned into three substructures, and the elements in the substructures are labeled in **Figure 3(c)**. Accordingly, there are 17 updating parameters in the first substructure, 15 in the second, and 13 in the third. The recorded input and output time history were analyzed in Matlab platform to derive the first 14 experimental frequencies and mode shapes.

Using the forward substructuring method, the first 30 modes in each substructure are selected as the master modes. In the model updating process, the substructure-based eigensolutions are compared with the first 14 experimental frequencies and mode shapes to form the objective function. The eigensensitivities are computed from one substructure solely to improve the computational efficiency. The elemental parameters of the FE model are adjusted iteratively to minimize the objective function through an optimal process. The elemental stiffness reduction factor (SRF) is used to estimate the damage identification, which gives the change ratio of the updated values to the initial values of updating parameters.

$$\mathrm{SRF} = \frac{\Delta\alpha}{\alpha} = \frac{\alpha^U - \alpha^O}{\alpha^O} \tag{70}$$

where superscript O denotes the initial values before updating and U denotes the updated values. The SRF values of the three substructures after updating are listed in **Figure 4(a)**. The model improved in the undamaged state is used for damage identification subsequently.

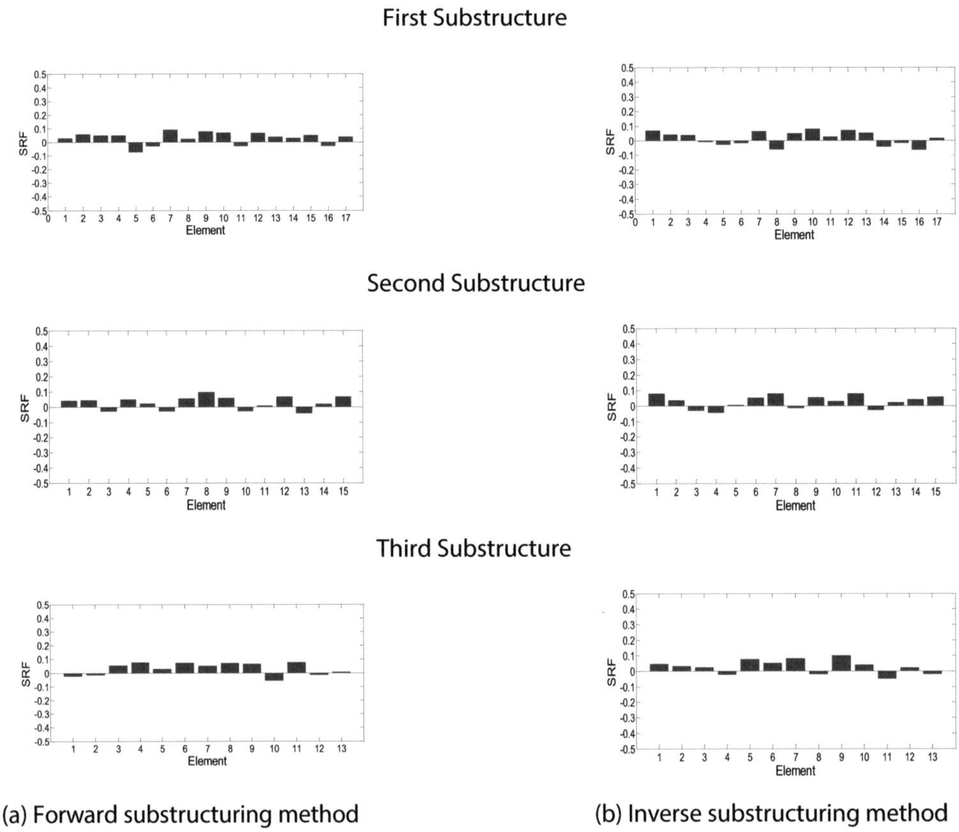

Figure 4. SRF values of the three substructures in the undamaged state. (a) Forward substructuring method. (b) Inverse substructuring method.

There are two damage configurations in the frame. In the first damage case, the column of the first storey is cut with the width of $b = 10$ mm and depth $d = 15$ mm at 180 mm away from the support (**Figure 3(b)**). Subsequently, the second storey is cut with the same width and depth at 750 mm away from the support.

In the first damage configuration, the cut is located in the first storey. The 17 elemental parameters in Substructure 1 are adjusted iteratively to minimize the discrepancy between the analytical eigensolutions and the measured modal data. In FE model updating, only the first substructure is reanalyzed, and the eigensolutions of the second and third substructures remain untouched and reused directly to compute the eigensolutions of global structure. The eigensensitivities with respect to the 17 elemental parameters are computed from the substructural derivative matrices of the first substructure solely, whereas those in the second and third substructures are zero-matrices. The elemental parameters in the undamaged state are subsequently employed for damage identification. It is apparent from **Figure 5(a)** that, Element 2 has an obvious negative value in SRF of about −25%, which agrees with the location of the cut in the experiment.

First Substructure

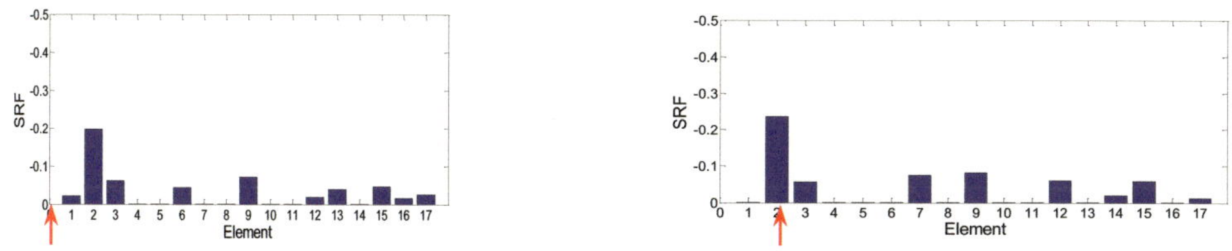

(a) Forward substructuring method

(b) Inverse substructuring method

Figure 5. SRF values of the first damage configuration. (a) Forward substructuring method. (b) Inverse substructuring method. ↑ Actual damage location.

In the second damage configuration, the two cuts are located in the first and second substructures, respectively. Subsequently, the first and second substructures are updated, while the third substructure remains untouched. The SRF values shown in **Figure 6(a)** demonstrate that Element 2 of the first substructure and Element 2 of the second substructure have an obvious negative SRF values. The identified locations agree with those of the experimental cut. Particularly, the SRF values of Element 2 of the first substructure are about −23%, comparable to that in the first damage configuration. This is because the cut remains unchanged in the two damage configurations.

Afterwards, the frame structure is analyzed by the inverse substructuring method with the same measured data and FE model. In the undamaged state, the global flexibility is formulated from the 14 pairs of measured natural frequencies and mode shapes. The inverse substructuring method is used to extract the substructural flexibility matrices of the three

First Substructure

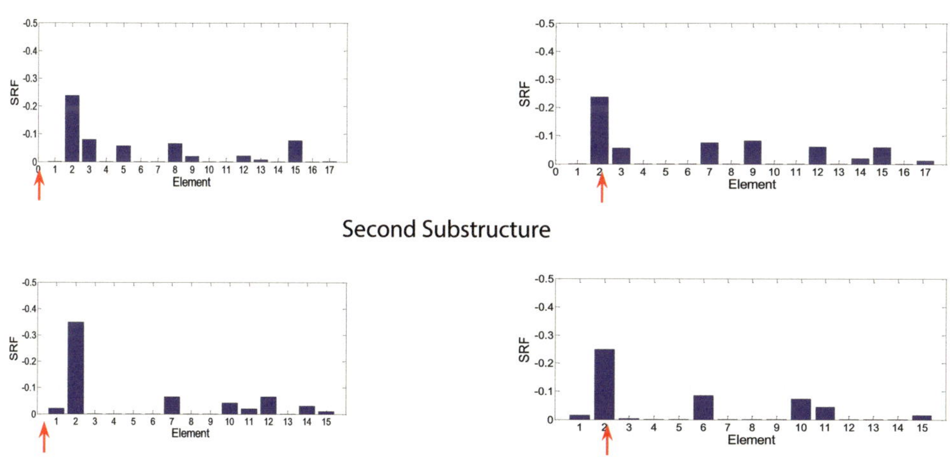

(a) Forward substructuring method

(b) Inverse substructuring method

Figure 6. SRF values of the second damage configuration. (a) Forward substructuring method. (b) Inverse substructuring method. ↑ Actual damage location.

substructures simultaneously. The global FE model is divided into three substructures as well. The substructural flexibility of the three submodels is compared with the extracted substructural flexibility to form the objective function. The discrepancy of substructural flexibility matrices between the FE sub-model and extracted ones is minimized by adjusting the updating elemental parameters of the three submodels independently. **Figure 4(b)** reports the updated SRF values of the three substructures, which are subsequently utilized for damage identification.

In the first damage case, the local area within the first storey, i.e., Nodes 1 to 18 in **Figure 3(c)**, are measured. Accordingly, only the substructural flexibility matrix of the first storey is extracted, based on which the submodel of the first substructure is updated independently. **Figure 5(b)** reveals a significant reduction in stiffness in Element 2, which agrees with the real location of the cut in experiment. The identified damage location and severity agrees with those obtained by the forward substructuring method as well.

In the second damage configuration, the frequencies and mode shapes measured in the first and second storeys are measured to form the global flexibility matrix. The substructural flexibility corresponding to the first and second substructures are extracted from the global flexibility simultaneously. The submodels of the first and second substructures are independently updated to recover the extracted substructural flexibility. **Figure 6(b)** reveals a negative SRF value of −20% in Element 2 of the first substructure and −25% in Element 2 of the second substructure. The identified damage location and severity are consistent to those gained by the forward substructuring method again. Both the forward and inverse substructuring methods are effective in model updating and damage identification.

5. Conclusion

A forward substructuring method and an inverse substructuring method are proposed in this chapter for model updating and damage identification. In the forward substructure-based model updating, the modified substructures are reanalyzed and assembled with other untouched substructures for the eigensolutions of the global structure to match the experimental data in an optimal manner. In the inverse substructuring method, the experimental modal data measured in local areas are used to extract the experimental flexibility matrix of the concerned substructure. The concerned substructures are updated by being treated as independent structures. Both the forward and inverse substructuring methods are effective in model updating and damage identification of a laboratory-tested steel frame structure. In the substructure-based model updating, only one substructure instead of the large-scale global structure is re-analyzed, which will be quite efficient for the model updating of practical large-scale structures. The substructuring methods are promising to be combined with the nonlinear analysis, vibration control, and parallel computation as well.

Author details

Shun Weng[1]*, Hong-Ping Zhu[1], Yong Xia[2] and Fei Gao[1]

*Address all correspondence to: wengshun@mail.hust.edu.cn

1 School of Civil Engineering and Mechanics, Huazhong University of Science and Technology, Wuhan, Hubei, P. R. China

2 Department of Civil and Structural Engineering, The Hong Kong Polytechnic University, Hung Hom, Kowloon, Hong Kong

References

[1] Mottershead JE, Friswell MI. Model updating in structural dynamics: a survey. Journal of Sound and Vibration. 1993;**167**:347–375. DOI: 10.1006/jsvi.1993.1340

[2] Bakira PG, Reynders E, Roeck GD. Sensitivity-based finite element model updating using constrained optimization with a trust region algorithm. Journal of Sound and Vibration. 2007;**305**: 211–225. DOI: 10.1016/j.jsv.2007.03.044

[3] Mottershead JE, Link M, Friswell MI. The sensitivity method in finite element model updating: a tutorial. Mechanical Systems and Signal Processing. 2011;**25**:2275–2296. DOI: 10.1016/j.ymssp.2010.10.012

[4] Klerk D, Rixen DJ, Voormeeren SN. General framework for dynamic substructuring: history, review, and classification of techniques. AIAA Journal. 2008;**46**:1169–1181. DOI: 10.2514/1.33274

[5] Weng S, Xia Y, Xu YL, Zhou XQ, Zhu HP. Improved substructuring method for eigensolutions of large-scale structures. Journal of Sound and Vibration. 2009;**323**:718–736. DOI: 10.1016/j.jsv.2009.01.015

[6] Weng S, Zhu HP, Xia Y, Mao L. Damage detection using the eigenparameter decomposition of substructural flexibility matrix. Mechanical Systems and Signal Processing. 2013;**34**:19–38. DOI: 10.1016/j.ymssp.2012.08.001

[7] Weng S, Xia Y, Xu YL, Zhu HP. Substructure based approach to finite element model updating. Computers and Structures. 2011;**89**:772–782. DOI: 10.1016/j.compstruc.2011.02.004

[8] Choi D, Kim H, Cho M. Iterative method for dynamic condensation combined with substructuring scheme. Journal of Sound and Vibration. 2008;**317**:199–218. DOI: 10.1016/j.jsv.2008.02.046

[9] Weng S, Zhu HP, Xia Y, Zhou XQ, Mao L. Substructuring approach to the calculation of higher-order eigensensitivity. Computers and Structures. 2013;**117**:23–33. DOI: 10.1016/j.compstruc.2012.11.005

[10] Bathe KJ, Wilson EL. Numerical Methods in Finite Element Analysis. Prentice-Hall Inc. Englewood chiffs, New Jersey: Wiley; 1989. DOI: 10.1002/nag.1610010308

[11] Sehmi NS. Large order structural eigenanalysis techniques algorithms for finite element systems. England, Ellis Horwood Limited, Chichester: Wiley; 1989. DOI: 10.1002/cnm.1630060609

[12] Fox RL, Kapoor MP. Rate of change of eigenvalues and eigenvectors. AIAA Journal. 1968;**6**:2426–2429.

[13] Nelson RB. Simplified calculation of eigenvector derivatives. AIAA Journal 1976;**14**:1201–1205.

[14] Perera R, Ruiz A. A multistage FE updating procedure for damage identification in large-scale structures based on multiobjective evolutionary optimization. Mechanical Systems and Signal Processing. 2008;**22**:970–991. DOI: 10.1016/j.ymssp.2007.10.004

[15] Ewins DJ. Modal Testing: Theory, Practice and Application. 2th ed. Baldock, England: Research Studies Press; 2000.

[16] Alvin KF, Park KC. Extraction of substructural flexibility from global frequencies and mode shapes. AIAA Journal. 1999;**37**:1444–1451. DOI: 10.2514/2.621

[17] Doebling SW, Peterson LD. Experimental determination of local structural stiffness by disassembly of measured flexibility matrices. Journal of Vibration and Acoustics. 1998; **120**:949–957. DOI: 10.1115/1.2893925

[18] Felippa CA, Park KC, Filho MRJ. The construction of free-free flexibility matrices as generalized stiffness inverses. Computers & Structures. 1998;**68**:411–408. DOI: 10.1016/S0045-7949(98)00068-6

[19] Jaishi B, Ren WX. Damage detection by finite element model updating using modal flexibility residual. Journal of Sound and Vibration. 2006;**290**:369–387. DOI: 10.1016/j.jsv.2005.04.006

[20] Guyan RJ. Reduction of stiffness and mass matrices. AIAA Journal. 1965;**3**:380. DOI:10.2514/3.2874

[21] Xia Y, Lin RM. Improvement on the iterated IRS method for structural eigensolutions. Journal of Sound and Vibration. 2004;**270**:713–727. DOI: 10.1016/S0022-460X(03)00188-3

[22] Xia Y, Lin RM. A new iterative order reduction (IOR) method for eigensolutions of large structures. International Journal for Numerical Method in Engineering. 2004;**59**:153–172. DOI: 10.1002/nme.876

[23] Formenti DL, Richardson MH. Parameter estimation from frequency response measurements using rational fraction polynomials. In: The 20th International Modal Analysis Conference; 04–07 February 2002; Los Angeles, CA; 2002. pp. 1–10.

Chapter 2

A New Method of SHM for Steel Wire Rope and its Apparatus

Shiwei Liu, Yanhua Sun and Wenjia Ma

Additional information is available at the end of the chapter

http://dx.doi.org/10.5772/intechopen.68148

Abstract

Steel wire ropes often operate in a high-speed swing status in practical engineering, and the reliable structural health monitoring (SHM) for them directly relates to human lives; however, they are usually beyond the capability of present portable magnet magnetic flux leakage (MFL) sensors based on yoke magnetic method due to its strong magnetic force and large weight. Unlike the yoke method, a new method of SHM for steel wire rope is proposed by theoretical analyses and also verified by finite element method (FEM) and experiments, which features much weaker magnetic interaction force and similar magnetization capability compared to the traditional yoke method. Meanwhile, the relevant detection apparatus or sensor is designed by simulation optimization. Furthermore, experimental comparisons between the new and yoke sensors for steel wire rope inspection are also conducted, which successfully confirm the characterization of smaller magnetic interaction force, less wear, and damage in contrast with traditional technologies. Finally, methods for SHM of steel wire rope and apparatus are discussed, which demonstrate the good practicability for SHM of steel wire rope under poor working conditions.

Keywords: steel wire rope, open magnetizer, structural health monitoring (SHM), notch filter, wavelet denoising

1. Introduction

Steel wire ropes are vital components concerning life safety and production, which usually operate in high-speed swaying conditions, therefore urgently being demanded to be monitored by nondestructive testing (NDT) methods for their structural health [1–8]. In most cases such as elevator, hoister in construction site, or in the mine, steel wire ropes are often driven by a friction wheel and run in turning motion with the aid of guide wheel, and then the dragging cages move up and down in a common working site. Owing to the spindle and soft features of the ropes themselves (max length: 3000 m), high-speed (max: 15 m/s) swaying and nonuniform of loading, random shaking frequently happens to the operational ropes with a large distance (max: 50 mm) from detection sensors. What is more, due to poor service, various forms of steel wire rope failure such as corrosion, fracture, and fatigue wear frequently occur. Consequently, structural healthy monitoring (SHM) for steel wire rope plays a significant role in ensuring security and preserving assets.

As the detection apparatus or sensors, although some portable detection devices based on magnet yoke configuration have been developed, their magnetization and relevant detection depth are limited due to the limited permanent-magnet yoke configuration number and size. In view of these abovementioned, new SHM sensor and apparatus based on electric-magnetic flux leakage (MFL) detectors having adjustable and sufficient magnetizing strength by adjusting electric current compared to the permanent magnet-MFL ones for small radial steel rope [9] is proposed, which has powerful detection abilities, especially for the internal defects in steel wire ropes. More importantly, they are usually the first choice for high-speed MFL testing apparatuses for steel wire rope with large radius. Therefore, electromagnetic methods suitable for highly efficient SHM (fast detection) and antipollution are introduced.

Finally, owing to that steel wire ropes are composed of several strands of steel wires; another challenge of SHM for them is the signal classification or differentiating. Therefore, signal-processing methods are also discussed.

2. New SHM sensor and apparatus

2.1. SHM sensor and apparatus for the new proposed method

An electrical wire was discovered to generate magnetic fields by Hans Christain Oersted in 1820; as shown in **Figure 1(a)**, an electrical wire in a C-shape [10] will produce magnetic fields in the internal area of the arc. However, when applying electricity to the C-shaped wire shown in **Figure 1(b)**, the electrical current has both forward and backward directions. Accordingly, to further promote the magnetic-gathering actions of a carrying-current arc, we connect the arc to form a loop that is then formed into a C-shape configuration with an opening, as shown in **Figure 1(b)**. The magnetic fields at the arc center are theoretically very weak due to the offset of forward current (I) and backward current (I'), which can be calculated as follows:

$$B = B_1 - B_2 = \int_0^{\varphi} d(B_1 - B_2) = \int_0^{\varphi} dB_1 - \int_0^{\varphi} dB_2 = \left| \frac{\mu_0 I \varphi}{4\pi r_1} - \frac{\mu_0 I' \varphi}{4\pi r_2} \right|_{r_1 \to r_2} \to 0 \qquad (1)$$

where μ_0 is the permeability of vacuum, φ is the angle of current element and radius vector, B_1 and B_2 represent the magnetic flux density produced by current I and I',r_1 and r_2 are the internal and external radii of the arc, respectively.

The magnetic fields inside the electrical arc are too weak to use for saturation magnetizing of the tested objects, which mainly arises from the offset of the "positive" and "negative" magnetic fields separately produced by the "forward" current (I) and "backward" current (I'), as shown in **Figure 1(c)** [11]. As a consequence, a ferrous core is arranged between the forward

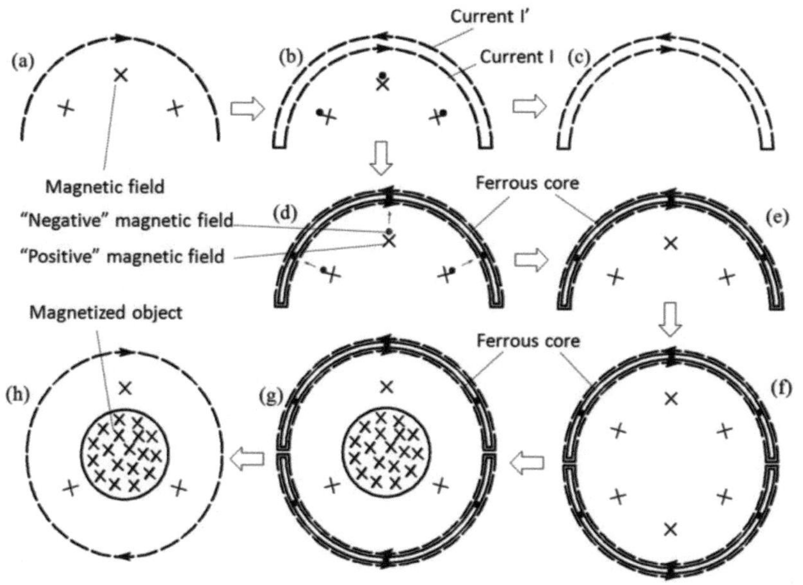

Figure 1. The magnetizing principle of an open coil. (a) Electrical wire in C-shape; (b) A loop with C shaped configuration; (c) Magnetic field inside the electrical arc; (d) Electrical wire in C-shape with a ferrous core; (e) Magnetic field inside the electrical arc with ferrous core; (f) Open coils with ferrous core; (g) Magnetic field produced by open coils in the magnetized object; (h) Magnetic field produced by conventional cylindrical solenoid in the magnetized object.

current (I) and the backward current (I'), as shown in **Figure 1(d)**; this core is designed to increase the "positive" magnetic fields (H) of the forward current using the magnetic-gathering effect. In addition, the ferrous core attracts the "negative" magnetic fields of the backward current (I') toward itself and removes this negative field from the arc center. By eliminating the negative effects caused by the backward current, the magnetic fields return to the value described in **Figure 1(a)**, as shown in **Figure 1(e)**. Finally, a magnetic field effect similar to the traditional electric annular loop can be obtained if the open coils are embedded with a ferrous core that consists of two identical half-open coils, as shown in **Figure 1(f)–(h)**. Finally, this configuration should produce a magnetic energy conversion region similar to a conventional cylindrical solenoid, as shown in **Figure 1(h)**.

Accordingly, a new open electric magnetizer is proposed, as shown in **Figure 2**, which consists of a C-shaped loop coil and a C-shaped ferrous core. This magnetizer primarily features an open mouth capable of encircling objects unlike the traditional tubular solenoid used as a magnetizer. Functionally, the proposed open magnetizer can center on an elongated object, such as in-service mine hoist wire ropes, to supply them with magnetization, which is beyond the capability of a conventional cylindrical electric magnetizer (i.e., solenoid). The advantage for elongated objects is illustrated in **Figure 2**.

The current electromagnetic methods failed to detect steel wire rope structures without heads and tails due to the tubular solenoids used as electro-magnetizers. Accordingly, a new SHM method based on the open magnetizer for steel wire rope without ends is proposed here. As schematically illustrated in **Figure 3**, the testing equipment consists of an open test probe, a direct current (DC) power supply, relevant data acquisition (DAQ) modules, and display

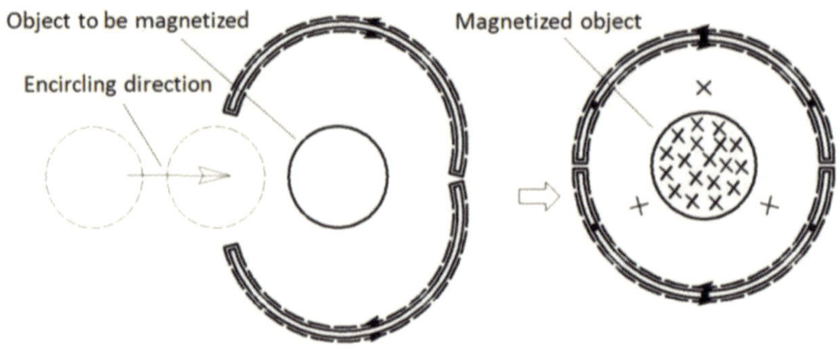

Figure 2. The proposed open electric magnetizer.

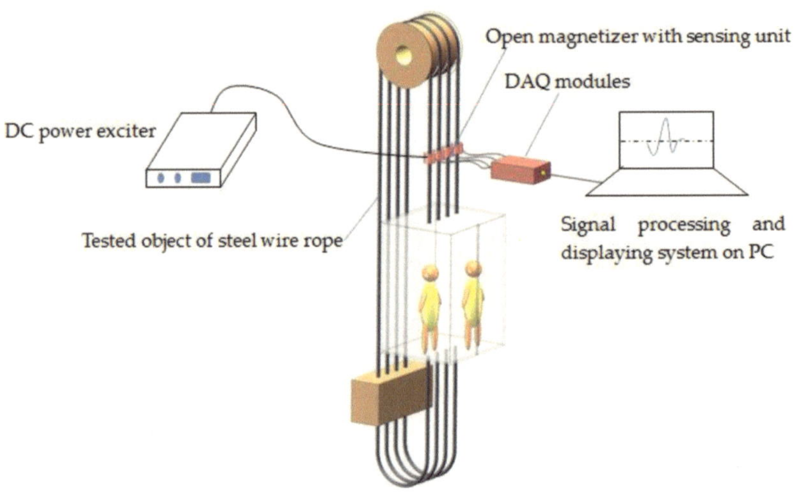

Figure 3. SHM methodology for steel wire rope using the open magnetizer.

software. The open test probe consists of an open magnetizer with magnetic-sensing units. The test procedure is as follows:

I. Encircle the steel wire rope in the center of the open magnetizer with a magnetic sensing unit. II. Apply DC power excitation to the open magnetizer coils and magnetize the steel wire rope. III. Scan the steel wire rope along the axial direction to acquire data and detect defects by moving the open magnetizer coils. IV. Observe and judge the signals from the displaying software on a personal computer (PC) to detect any defects.

An ANSYS simulation, as shown in **Figure 4(a)**, was built to describe a section of the open magnetizer and steel rebars; the internal diameter is 60 mm, the ferrous half C-shaped core is 15 mm thick, and the axial lengths of both the core and the coil are 120 mm. The tubular electromagnetic magnetizer consisting of the tubular coil has an internal diameter of 60 mm, a thickness of 15 mm in the radial direction, and an axial length of 120 mm, as displayed in **Figure 4(b)**. As indicated in **Figure 4(c)**, the specimen to be magnetized is a group of steel bars

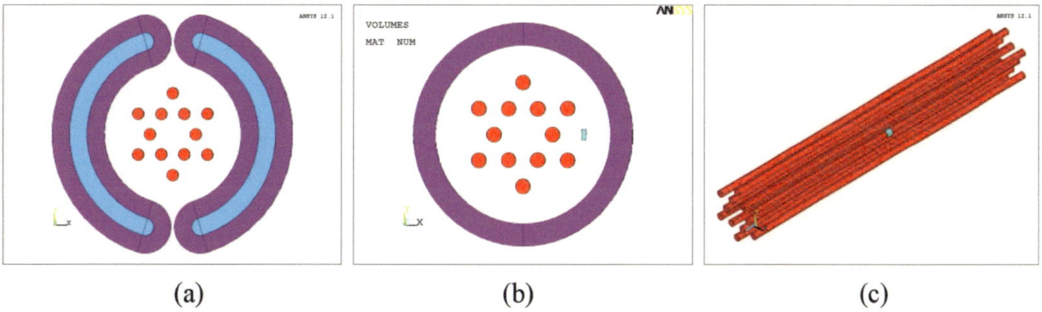

Figure 4. Three-dimensional models in the ANSYS simulations: (a) the sectional model of the open coil with the steel wire ropes; (b) the sectional model of the tubular coil with steel wire ropes; and (c) the model of the steel rope specimens with defects.

that have an axial length of 500 mm and a diameter of 10 mm; the axial distance between the outer steel wire rope and the magnetizer is 35 mm. Six steel wire ropes with defects in the axial direction were arranged in either the external layer or the internal layer at a 60° central angle. The Solid 117 element [12, 13] was chosen to build the three-dimensional models, and the magnetic permeability of the ferrous core was selected based on a B-H curve. Both sweep and free mesh methods [14, 15] were used to reduce the mesh number and to increase the symmetry of the mesh. For the mesh surrounding the defects in the steel ropes, the detailed mesh method was conducted to obtain more accurate calculations for the defect signals. The extracting operation was conducted in the route/path with an axial length of 120 mm and a diameter of 40.5 mm in the radial direction. The electrical current density was $J = 1e7$ A/m^2 for both the open coil and the tubular coil. The open coil model was separated into four circular arcs that each have cylindrical coordinates to separately apply the electrical current density. The solution method was selected to be difference scalar potential, and the boundary condition is Az = 0. The extracting path for the signal is in the axial direction of the cylindrical coordinates, with a starting point of (41e−3, 0, −50e−3) and an ending point of (41e−3, 0, 50e−3).

Using the 3D models mentioned above, the magnetization function and effect, defect detection capability and magnetic force interaction of an open magnetizer coil and a tubular coil are compared. As shown in **Figure 5**, the whole magnetization status produced by the 3D models

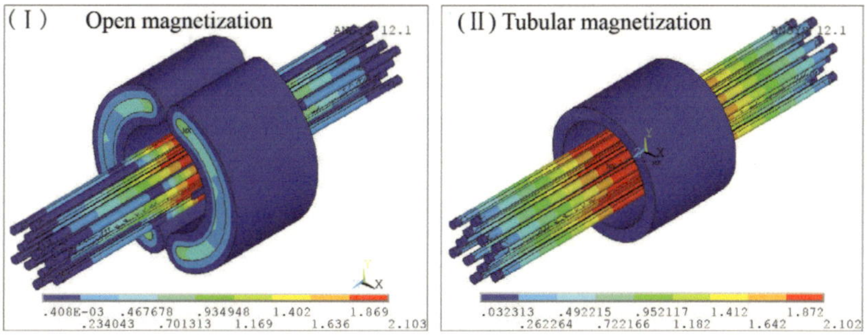

Figure 5. Magnetization status produced by the 3D models.

was obtained. The magnetic field density in both the radial and axial directions in the internal zone of the coils is displayed in **Figures 6** and **7**. As shown in **Figure 6**, the open magnetizer has a stronger radial magnetic field density, especially in the area close to the inner coils, than the tubular coils. **Figure 7** shows that the open magnetizer has a magnetization function and effect that is similar to that of the tubular one.

For the defect detection capability, the signals presenting the magnetic flux leakage density of both the internal and external defects in the steel wire ropes were obtained by path operation in the simulations, as shown in **Figure 8**. In this figure, "Open Inner" and "Open Outer" represent the signals excited by the open magnetizer coil, and "Traditional Inner" and "Traditional Outer" represent the signals excited by the traditional tubular magnetizer coil. Based on the simulation results shown in **Figure 8**, the open electromagnetic coil and the traditional tubular one have similar defect detection capabilities.

Figure 9 compares the magnetic force and shows that both the open magnetizer and the tubular one have approximate magnetic force interaction with the steel wire ropes. Therefore, these methods are suitable for defect detection, unlike permanent magnet magnetizers.

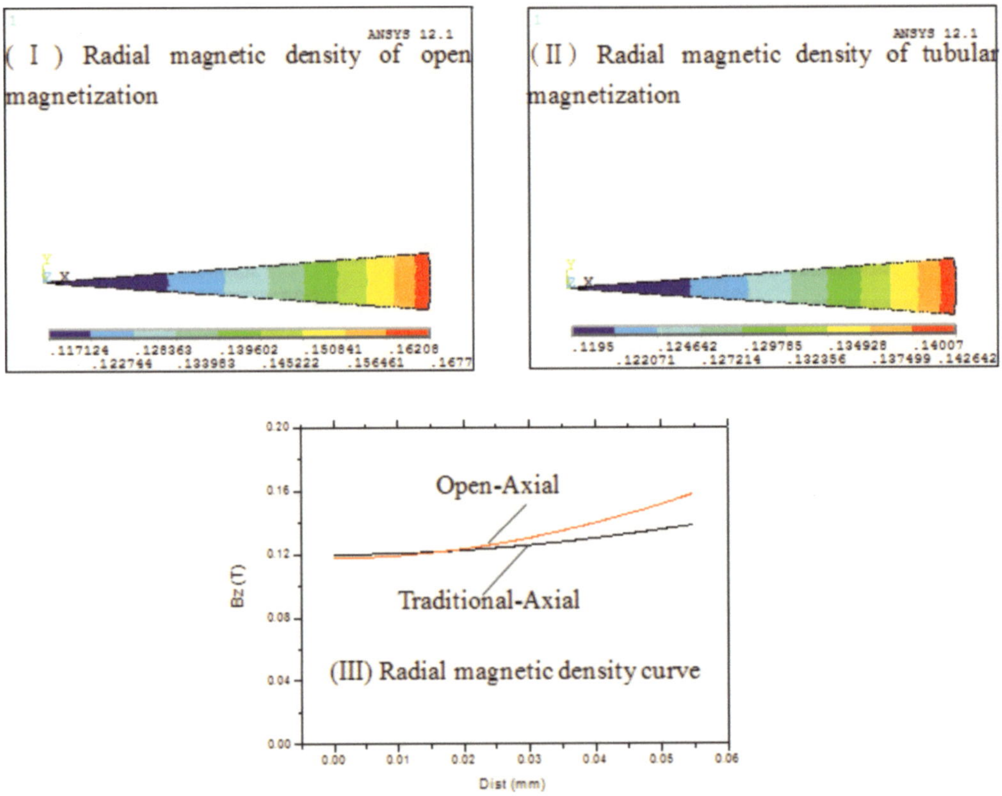

Figure 6. Magnetic field density in the radial direction inside the coils.

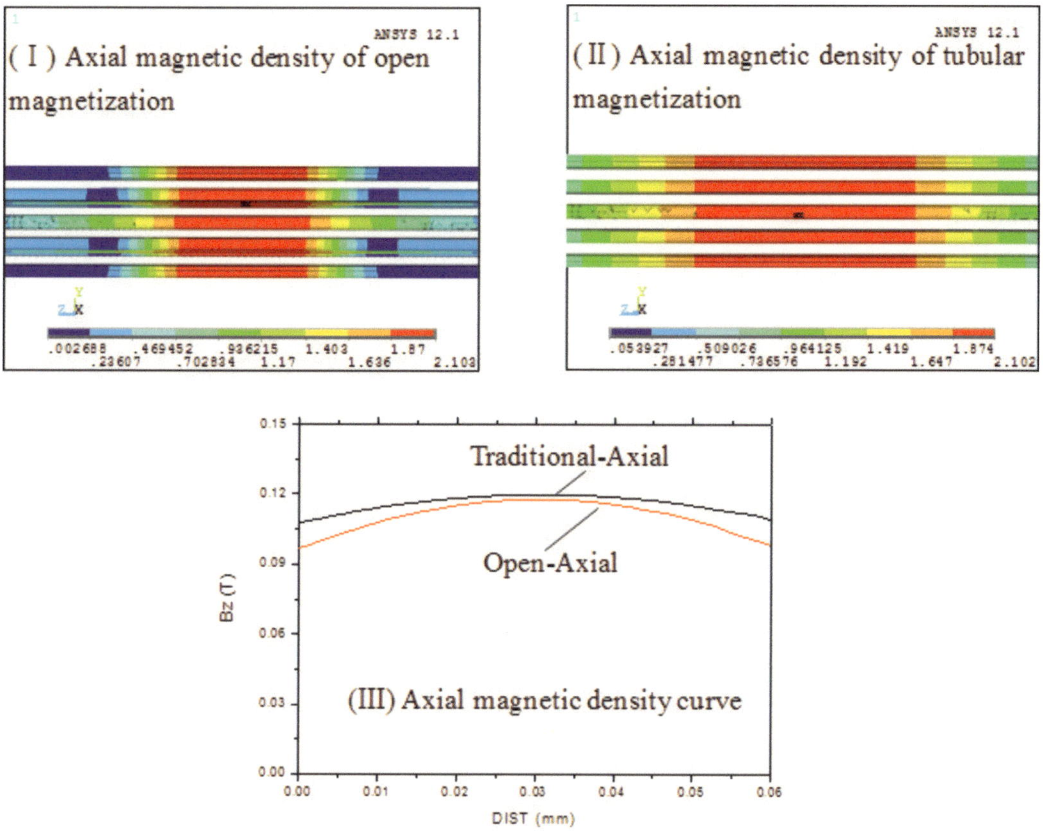

Figure 7. Magnetic field density in the axial direction inside the coils.

After finding similar magnetization effects and detect detection capabilities for both the proposed open electromagnetic coil and the traditional tubular one from the previous simulations, the open coil was further optimized. As shown in **Figure 10**, the relationships between the magnetization capability and the radial thickness, magnetic permeability and axial length of the C-shaped core were obtained. **Figures 10(a)–(c)** show that the optimum core is 15 mm in radial thickness, made of 45# steel and 120 mm in axial length.

Based on the previous analyses and optimization simulations, the traditional tubular coil and the open electromagnetic coil with a ferromagnetic core inside were designed and manufactured, as shown in **Figure 11**. The traditional tubular coil was mainly composed of copper wire wound outside the rubber mold. To verify the reliability and the validity of the simulations, optimized sizes for the magnetizer coils were chosen based on the analysis above. As illustrated in **Figures 11(a)** and **(c)**, the axial lengths of both magnetizer coils are 120 mm, the thickness of the open coil is 15 mm, and the material of the open electromagnetic core is 45# steel. The inner diameters of both magnetizer coils are 130 mm, based on the outer diameters of

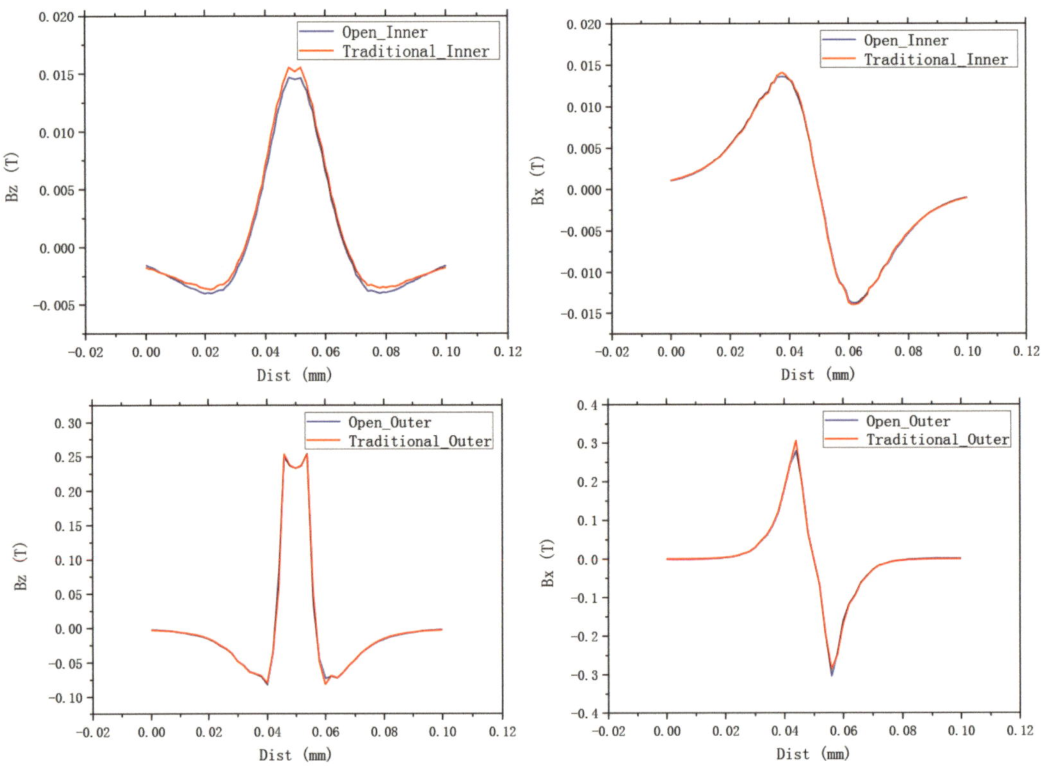

Figure 8. Comparison of the detection capabilities of the open and tubular magnetizers.

the tested concrete rebar, and are adjustable. Pictures of the actual electromagnetic coils are shown in **Figures 11(b)** and **(d)**.

First, a magnetization function experiment was conducted to measure the magnetic density using a Gauss meter, as shown in **Figure 12**. According to the simulations, both the radial and axial directions were tested, as shown in **Figure 12(a)**; **Figure 12(b)** shows the actual detection apparatus for these magnetizer coils.

These electromagnetic coils were magnetized using a DC source of 3A, and all of the magnetization data produced by the electromagnetic magnetizer were recorded by moving the magnetic sensor probe of the Gauss meter to different locations along the indicated radial and axial directions. The magnetic field densities in both the radial and axial directions inside the coils are displayed in **Figure 13(a)** and **(b)**; the coordinate "DIST" is the distance of the probe from the center of the coils. **Figure 13(a)** shows that the open magnetizer has stronger radial magnetic field density than the tubular coil, especially in the area close to the inner coil. As shown in **Figure 13(b)**, the open magnetizer has a similar magnetization function as the tubular one. All of the measurement results are in accordance with the simulation results.

Furthermore, an experimental apparatus simulating detection was built. After the experiments were conducted, the detection signals for the man-made defects were saved, as shown in **Figure 12**. Using different directions of the hall element sensor inside the magnetizing coils,

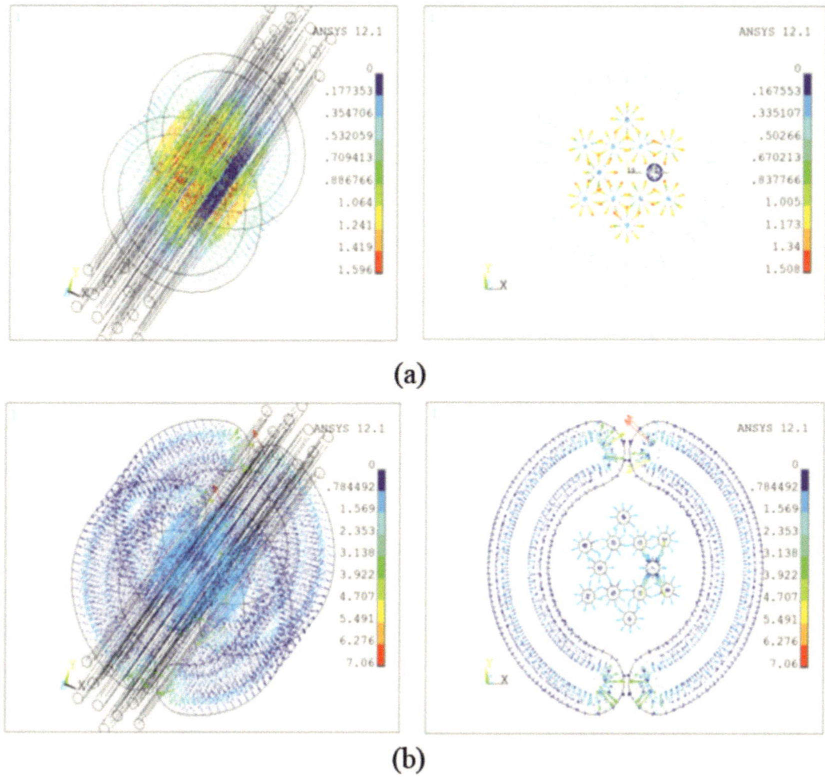

Figure 9. Comparison of magnetic force of tubular and open magnetizers. Magnetic force interaction of (a) tubular coils and (b) open coils.

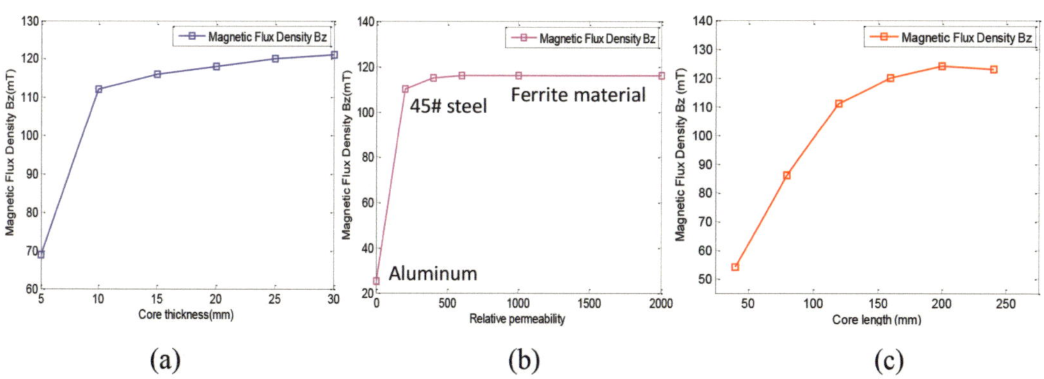

Figure 10. Relationships between the magnetization capability and the structural parameters of the core: (a) relationship between the magnetization capability and the radial thickness of the core; (b) relationship between the magnetization capability and the magnetic permeability of the core; and (c) relationship between the magnetization capability and the axial length of the core.

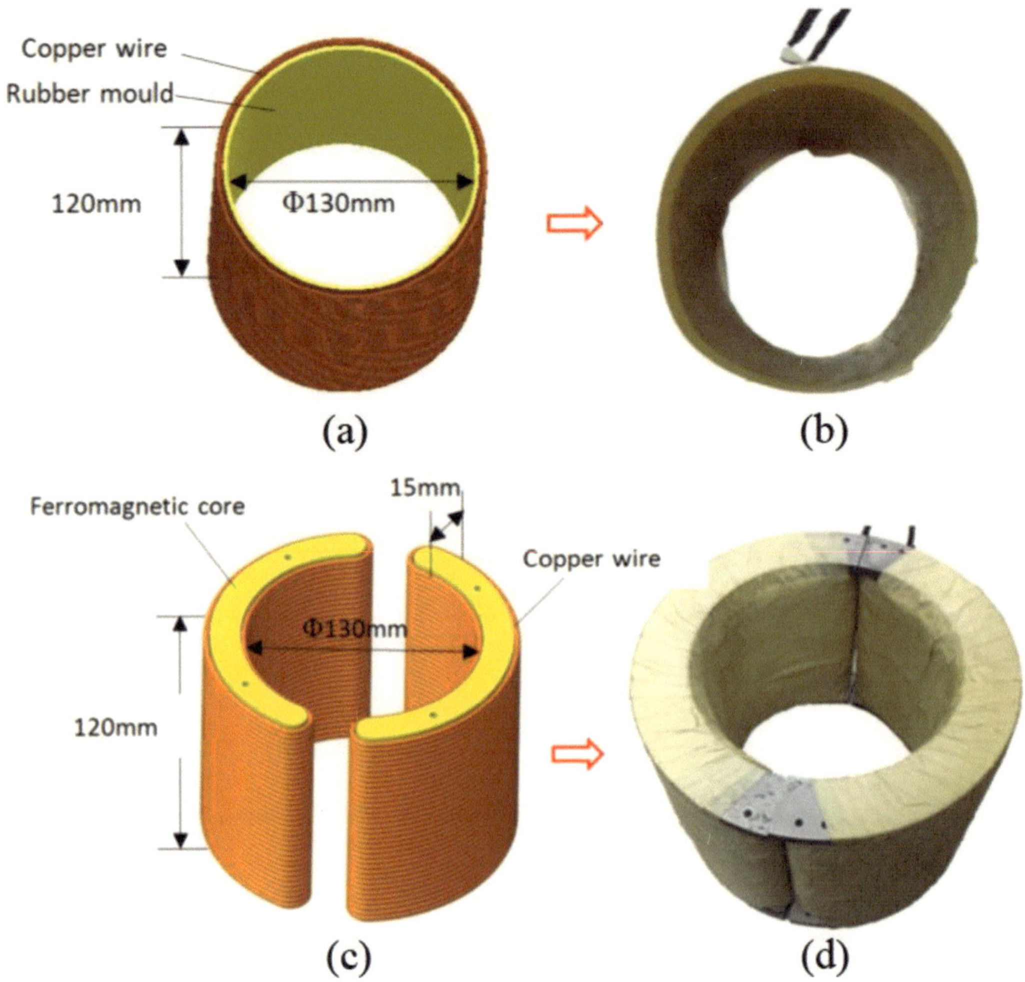

Figure 11. Traditional tubular coil and the proposed open electromagnetic coil. (a) Schematic diagram of the tubular coil; (b) actual tubular coil; (c) schematic diagram of the open coil and (d) actual open electromagnetic coil.

the axial and radial amplitudes of the MFL signals were obtained. The used 3551-Hall-array sensor is located closely to the internal wall of the open coil at the center of the coil in its axial direction. The moving speed of the detection probe for scanning is about 0.25 m/s with respect to the concrete rebar.

According to the signals shown in **Figure 14**, "Traditional" represents signals that were acquired by the traditional magnetizing tubular coil, as mentioned in **Figure 11(b)**, while "Open" means that the signals were obtained using the novel open electromagnetic coil, as described in **Figure 11(d)**. Similarly, "Inner" and "Outer" denote that the signals were produced by defects in the internal layer and the external layer, respectively.

As shown in **Figure 14(a)** and **(c)**, both magnetizing coils have a strong ability to detect internal defects; in addition, the defect signals from the external layer are stronger than those of the

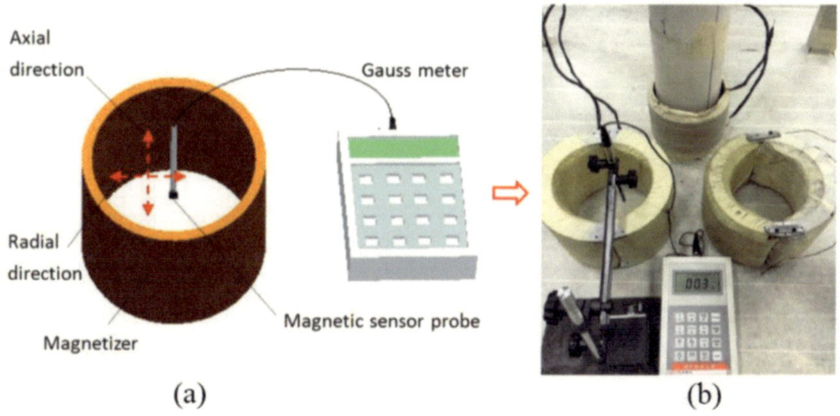

Figure 12. Magnetic density detection experiment. (a) Magnetic density detection using a Gauss meter and (b) actual detection apparatus.

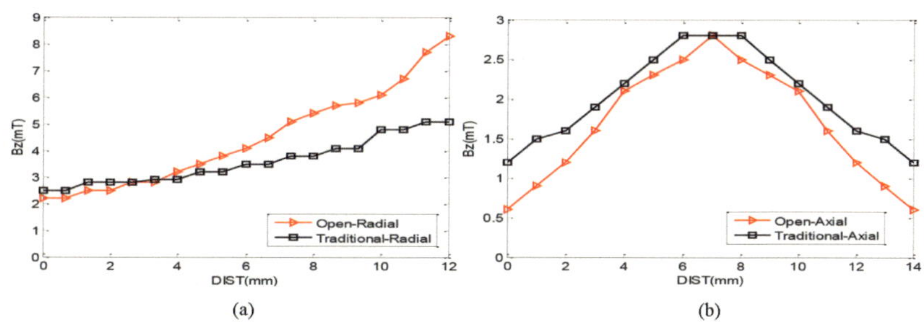

Figure 13. Magnetic density measured using the Gauss meter. (a) Radial magnetic density of the open and traditional coils measured using the Gauss meter and (b) axial magnetic density of the open and traditional coils measured using the Gauss meter.

internal layer. Moreover, the signals excited by the open electromagnetic coils are very similar to those of the traditional tubular coils. Furthermore, the experimental results are consistent with the simulation results shown in **Figure 8**, demonstrating the validity of the open electromagnetic technique proposed in this chapter.

2.2. Comparison of magnetic open and yoke MFL testing method

To address the hoist wire rope inspection challenges, this chapter also presents an open magnetization sensor possessing the advantages over yoke magnetization method [16–20] in terms of magnetic excitation capability and magnetic interaction force. Furthermore, this chapter also intends to design a relevant detection sensor based on open permanent magnetization method by optimization and a testing probe configuration capable of tracking swing rope by experimental comparison of MFL signal change trends. Finally, the chapter developed a corresponding portable testing apparatus with great promise of application performance in the future.

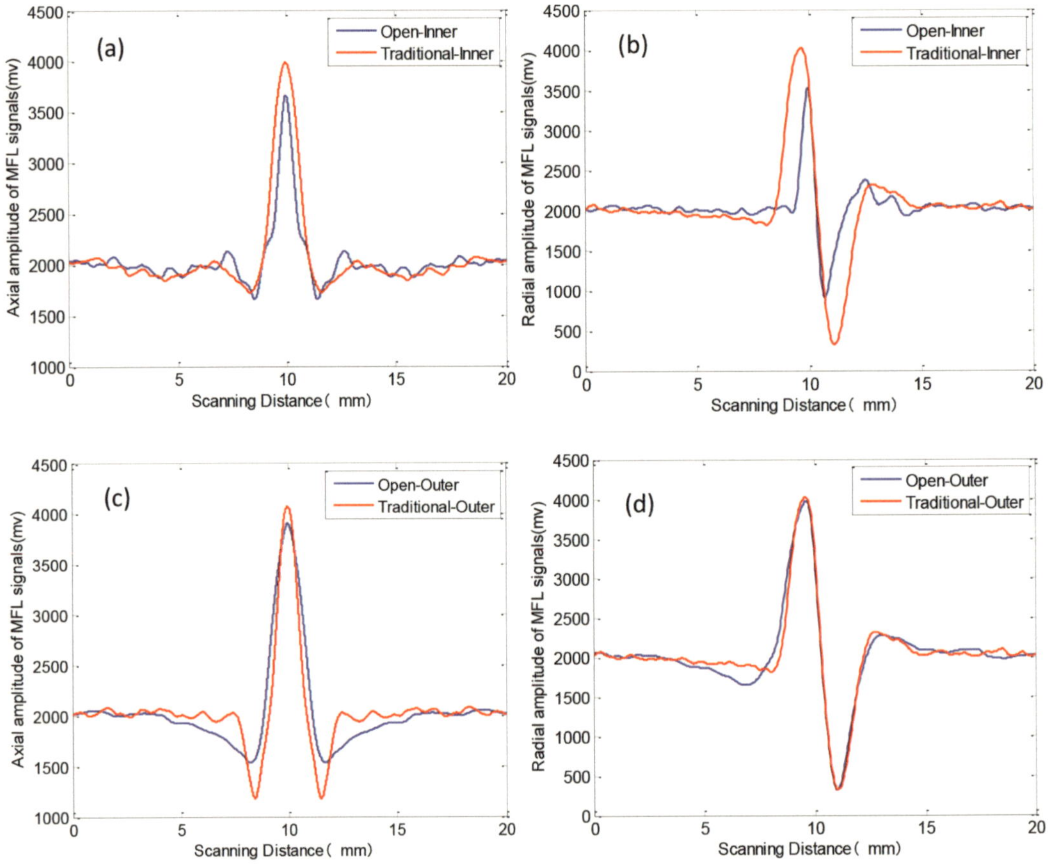

Figure 14. Experimental results of the defect detection capability of the two magnetizer coils; (a) Axial MFL signals by open and traditional coils for inner layer defects detection; (b) Radial MFL signals by open and traditional coils for inner layer defects detection; (c) Axial MFL signals by open and traditional coils for outer layer defects detection; (d) Radial MFL signals by open and traditional coils for outer layer defects detection.

Presently, the commonly used portable MFL testing instrument for static wire rope defect detection mainly consists of several magnetic-yoke testing packages. Each of them is composed of double permanent magnets stacks with magnetic poles facing closed to the tested object, ferrous yoke connecting the stacks, and magnetic-sensing units fixed under the yoke and adjacent to the poles. The magnetic-yoke MFL testing method for wire rope is schematically illustrated in **Figure 15**. To magnetically saturate the object being detected, the double magnetic poles should be located closely to the tested wire rope as much as possible, together with the sensing units, leading to a small lift-off distance (1–5 mm) for each package relative to the tested wire rope. Additionally, the ferrous yoke also should be designed at a certain length (over 240 mm) and height (over 50 mm) to ensure a linear working range for some magnetic sensors such as hall element. In consideration of these above, the current yoke-type MFL testing sensor is characterized by strong magnetic force interaction (at least 300 N) with the ropes, large volume (at least 240 mm in length), large weight (at least 9 kg in weight), and poor adaptability for tested wire ropes with different diameters. Consequently, with respect to the

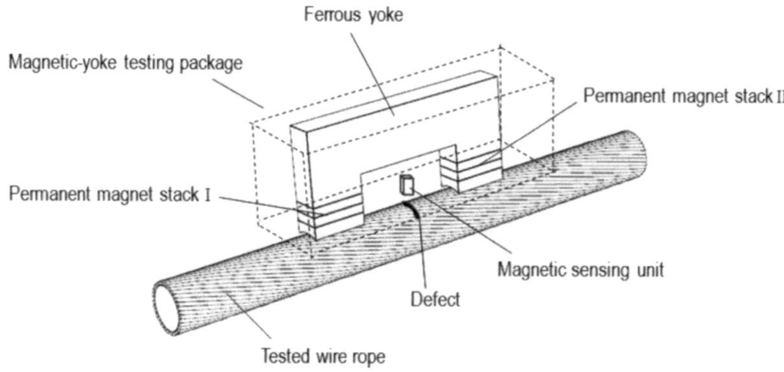

Figure 15. Testing principles based on yoke-magnetizing method for wire rope.

portable mine hoist wire ropes running in a high-speed-swaying status, the existing magnetic-yoke method is hard to perform the task of hoist wire rope detection.

Differing from the magnetic circuit principles mentioned above, an open magnetization method hopefully created by a ring-shaped magnet is here provided in contrast with the commonly used ring-shaped coils, which apply open magnetization technique to magnetize the detected objects. According to magnetic charge and molecular flow theory, the description for the magnetic potential ϕ_m at an arbitrary point of P outside the permanent is

$$\phi_m = \frac{1}{4\pi}\oint_S \frac{M \cdot n}{R} dS + \frac{1}{4\pi}\oint_\tau \frac{\nabla \cdot M}{R} d\tau \tag{2}$$

where M is the magnetic strength of permanent magnet, S is the curved surface encircling the magnet, R is the distance between the source point to the field point, n is normal to the external surface of magnet, and ∇ is divergence operator.

Owing to the uniform magnetization, the ring-shaped magnet manifests itself by surface current and features no body current. Therefore, there is equivalent electric current density for area of $\rho_{\mathrm{sm}} = M \cdot n$ and for volume of $\rho_m = \nabla \cdot M = 0$.

Thus, Eq. (2) can be

$$\phi_m = \frac{1}{4\pi}\oint_S \frac{\rho_{sm}}{R} dS \tag{3}$$

Then, combined with Obi-Wan thrall's law (i.e., $H = -\nabla \cdot \varphi_m$), magnetic flux density can be obtained by

$$H = \oint_S \frac{\rho_{\mathrm{sm}}}{4\pi} \cdot \frac{R}{R^3} dS \tag{4}$$

As shown in **Figure 16(a)**, the magnetic field density produced at an arbitrary point $P(\rho, \varphi, \mathrm{h})$ can be further calculated as

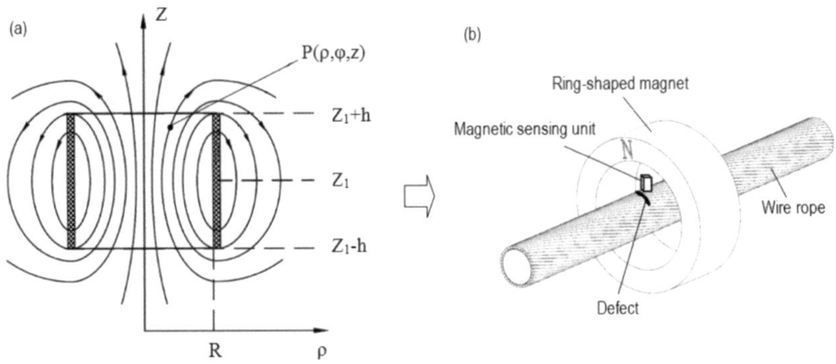

Figure 16. Testing principles based on open-magnetizing method; (a) Magnetic field density produced at an arbitrary point by the coil; (b) Ring shaped magnetizer in wire rope detection.

$$H(\rho,\varphi,h) = \frac{M}{4\pi}\iint_{S_+} \frac{(\rho-x)i+yj+(h-z)k}{\left((\rho-x)^2+y^2+(h-z)^2\right)^{3/2}}dS - \frac{M}{4\pi}\iint_{S_-} \frac{(\rho-x)i+yj+(h-z)k}{\left((\rho-x)^2+y^2+(h-z)^2\right)^{3/2}}dS \quad (5)$$

Furthermore, the magnetic field strength produced by the ring-shaped magnet can be obtained by vector sum, described as

$$H(\rho,\varphi,h)_{R-r} = H(\rho_R,\varphi,h) - H(\rho_r,\varphi,h) \quad (6)$$

Finally, the magnetic flux density in the magnetized body (relative magnetic permeability μ) provided by the open magnetization method can be obtained as

$$B(\rho,\varphi,h) = \mu \cdot H(\rho,\varphi,h) \quad (7)$$

Unlike the abovementioned conventional magnetic-yoke method, learning from the wide-spread solenoid coil as a magnetizer in fixed MFL apparatuses, a new ring-shaped MFL testing method is proposed here, as displayed in **Figure 16**. Here, a ring-shaped permanent magnet is adopted to axially magnetize the whole circumference of the tested wire rope at a certain length similar to a tubular solenoid coil does, and unattached magnetic-sensing units covering the full circumference of the tested rope are separately placed inside the ring magnet, as indicated in **Figure 16(a)**. Owing to the ring-shaped open magnetization configuration, the magnetic interactions between the magnetizer and wire rope are largely decreased. On account of that, the full circumference magnetizer is like a tubular coil with excitation current, as shown in **Figure 16(b)**; the ring-shaped magnetizer can be designed with a larger space inside for separately disposing the magnetic-sensing units with various sizes correspondingly to different wire ropes. As a consequence, the provided method owns the advantages of small magnetic force interactions with the tested ropes, simple configuration for realization, and all-purpose magnetizer characteristics for various-sized wire ropes.

2.3. Simulation results

Finite element method (FEM) was applied to simulate and analyze the comparisons, and the two above-presented magnetization methods were modeled for a same steel wire rope (rope diameter of 32 mm and length 300 mm, dextral rope strand number of 19) with a broken wire (defect diameter of 1.2 mm). The two models consist of a ring-shaped permanent magnet (axial length 80 mm, internal diameter 70 mm, and external diameter 100 mm) as an open magnetizer and six magnet yokes (yoke width 24 mm, yoke height 20 mm, yoke length 180, leg length 10 mm, leg spacing 100 mm, and yoke lift-off 8 mm) as a yoke magnetizer, respectively. In the former model, to ensure the versatility of an open magnetizer for different sized wire ropes, radial spacing should be somewhat larger and the size of 20 mm is selected here. By experience in practice application, a certain length (60 mm) of the middle of the ring-shaped magnet is replaced by a ferrous ring-shaped configuration with the same size either for the installation convenience of magnetic-sensing units or reducing cost. During the course of the simulations, the coercive force and relative magnetic permeability of permanent magnets made of NdFeB material were, respectively, 490,000 A/M and 1, and all the ferrous materials had the magnetic permeability based on a B-H curve. Finite element was chosen by Solid 236 and 3D 20-node element, which is applicable to 3D static electromagnetic analyses and to modeling air, iron, nonferrous materials, and permanent magnets. Additionally, the Solid 236 can be used with command macros such as "EMAGERR (electromagnetic relative error)" for calculating relative error and "EMFT (electromagnetic forces and torques)" for summarizing electromagnetic forces. Two finite element models were obtained by the usage of sweep mesh combined with free mesh for both high resolution and fast computation, as presented in **Figure 17(a)** and **(b)**, respectively.

After solving the finite element models designed above, magnetization results for the steel wire ropes were obtained, as displayed in **Figure 18**. In **Figure 18(a)** and **(b)**, it can be observed that both of the methods produce the similar magnetic flux density at the magnetizing center of the wire ropes and they are capable of carrying out similar magnetically saturated statuses for the objects. Therefore, we can draw a conclusion that the performance of the magnetic excitation provided by the open magnetization method for detecting defects is similar to that supplied by the traditional yoke one.

MFL sensors accomplish their defect scanning for the tested objects by dynamic contact and sliding friction. Magnetic interaction force between MFL sensors and tested objects has a great influence on the service life of both sensors and tested objects due to wear or impact damage. Additionally, great magnetic interaction force also affects the steady location in the center of the

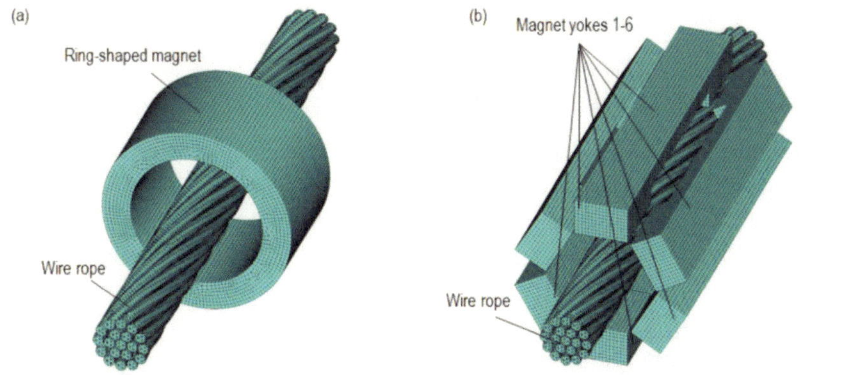

Figure 17. FEM models for open and yoke method; (a) FEM models for open method; (b) FEM models for yoke method.

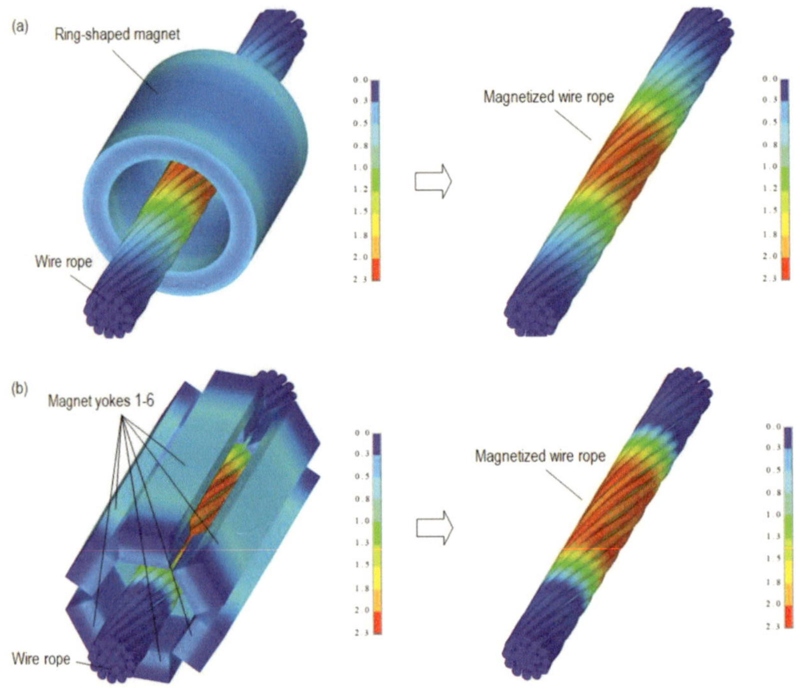

Figure 18. Magnetization capabilities for wire ropes, respectively, provided by open and yoke method; (a) Magnetization capabilities for wire ropes by open method; (b) Magnetization capabilities for wire ropes by yoke method.

magnetizer and the quick realization for detaching the sensor from the tested object. These problems are apparent in the portable inspections for the elongated objects in service. Therefore, the magnetic interaction force brought out by relevant methods should be concerned. Considering that MFL sensors are usually designed into two half parts to encircle the tested objects, the magnetic interaction force generated by the double half structures of both magnetization methods was extracted by vector plot of FMAG (magnetic force), as shown in **Figure 19**.

Furthermore, the detailed values of magnetic forces were calculated and summarized by directly using the command stream of FEM tool ANSYS of "EMFT," and the forces were stored as items F_x, F_y, F_z, and F_{total} parameters, as shown in **Table 1**.

The items of F_x, F_y, and F_z are the magnetic force component in Cartesian coordinate system listed in **Table 1**, and F_{total} is the vector sum magnetic force of these components.

Figure 19 and **Table 1** demonstrate that the magnetic interaction force produced by the yoke magnetization method is seven times stronger than that caused by the open one especially for the component of F_y and F_{total}. Hence, a conclusion can be drawn that the open magnetization method can realize detaching the sensor from the tested object more easily and meanwhile causes less wear and damage to them compared with the yoke one. Consequently, the open magnetization method is more suitable for portable detections of hoist wire rope and owns a much longer service life. The detailed distributions of the magnetic interaction forces to the wire ropes caused by the both magnetizers can be seen in **Table 2**, respectively.

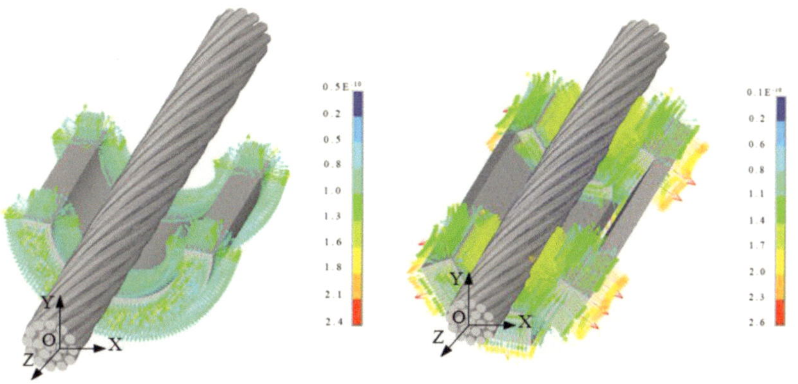

Figure 19. Magnetic interaction forces, respectively, caused by open and yoke model.

Model types	**Force types (N)**			
	F_x	F_y	F_z	F_{total}
Open method	0.01	45.30	0.0	45.30
Yoke method	0.02	360.01	0.0	360.01

Table 1. The summary results of the magnetic interaction force.

Model types	**Force types (N)**			
	F_x	F_y	F_z	F_{total}
Open method				
Yoke method		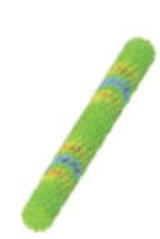		

Table 2. The detailed distributions of the magnetic interaction forces simulated by open and yoke method.

Regarding the primary design of open magnetization MFL sensor mentioned above, much attention should be focused on the ring-shaped permanent magnet magnetizer. Similar to the ring-shaped coil, the basic sizes of the ring-shaped permanent magnet sensors are axial length and the radial thickness. During the simulations, to maintain enough assembly space for the magnetic-sensing units, the distance between the surface of the tested wire ropes and the inner center of ring-shaped magnet always remained at 20 mm. The picking-up route of the magnetic field value (axial component of B_y and radial component of B_z) was set to be 20 mm in axial length and 1 mm in radial lift-off. Thereafter by changing relevant sizes of permanent magnet and picking up the magnetic fields via the specified route, the MFL signal waveforms produced by the broken defects of wire ropes were obtained as well as the curve of their peak-to-peak value, as demonstrated in **Figure 20**.

Similar to the commonly used tubular coils, the conclusion drawn from **Figure 20(a)** and **(b)** is that the larger axial length and radial thickness sizes that the ring-shaped magnet have, the stronger MFL signals would produce from wire rope defects when the axial length and radial thickness of the ring-shaped magnet are, respectively, less than 300 and 70 mm. Combed with the requirements of low cost in practice application as described previously, the middle part of the whole ring-shaped permanent magnet is creatively replaced by a ring-shaped ferromagnetic material while the two sides of the ferrous ring remain magnets. Therefore, **Figure 20(c)** and **(d)** were gained, which indicated that the MFL signals decrease with the increasing axial length of the ferromagnetic ring with a turning point of 80 mm and the MFL signal is the strongest when radial thickness of the ferromagnetic ring is 15 mm. Actually, permanent magnet ring was just defined by 15 mm according to practice experience at that time. Indeed, it has also been confirmed that the MFL is the strongest when the radial thickness of the ferromagnetic ring is the same as the permanent ring by additional simulations. As a result and combined with practice engineering application, the ferromagnetic ring is designed to be 80 mm in axial length and 15 mm in radial thickness, and the inner diameter is at least 70 mm to meet the testing requirement for various detected objects with different sizes.

2.4. Experimental results of the defect detection

Based on the simulation and size optimization done above, the open magnetization MFL sensor for wire ropes is designed, as displayed in **Figure 21**, which is simply composed of two parts, namely a ring-shaped magnet and its unattached ring-shaped shoe including magnetic-sensing units. The size specifications of ring-shaped shoes can be immediately obtained by changing its internal diameters, and one ring-shaped sensor can easily accomplish all the detections for different sized objects just through replacing the relevant size specifications of ring shoes. Explanatorily, the middle length of the ring magnet in its axial orientation is designed to be replaced by a ferromagnetic ring so that the unattached ring-shaped shoes could easily be fixed inside the inner ring and the whole cost could be reduced. Owing to that, the hoist wire ropes are fixed by carrying cages at both sides and have no head and tail, the designed open magnetizing sensor is easily designed into two-half structure in practice, which merely has the configuration sizes of 120 mm (length) × 100 mm (diameter) and just weighs about 1.5 kg, featuring small volume and weight.

As indicated in **Figure 22**, the open magnetization MFL sensor designed above was tested for the detection capability of a man-made broken-wire defect in a wire rope (Φ32 mm) and the magnetic force interaction with the tested wire rope. During the course of experimental testing for detection capability, the sensor was connected with an amplifier, a filter, an A/D converter, and a data analysis system in serials. By linear axial scanning along the rope, the output signals from the artificial broken wire were finally acquired, demonstrating that the proposed MFL

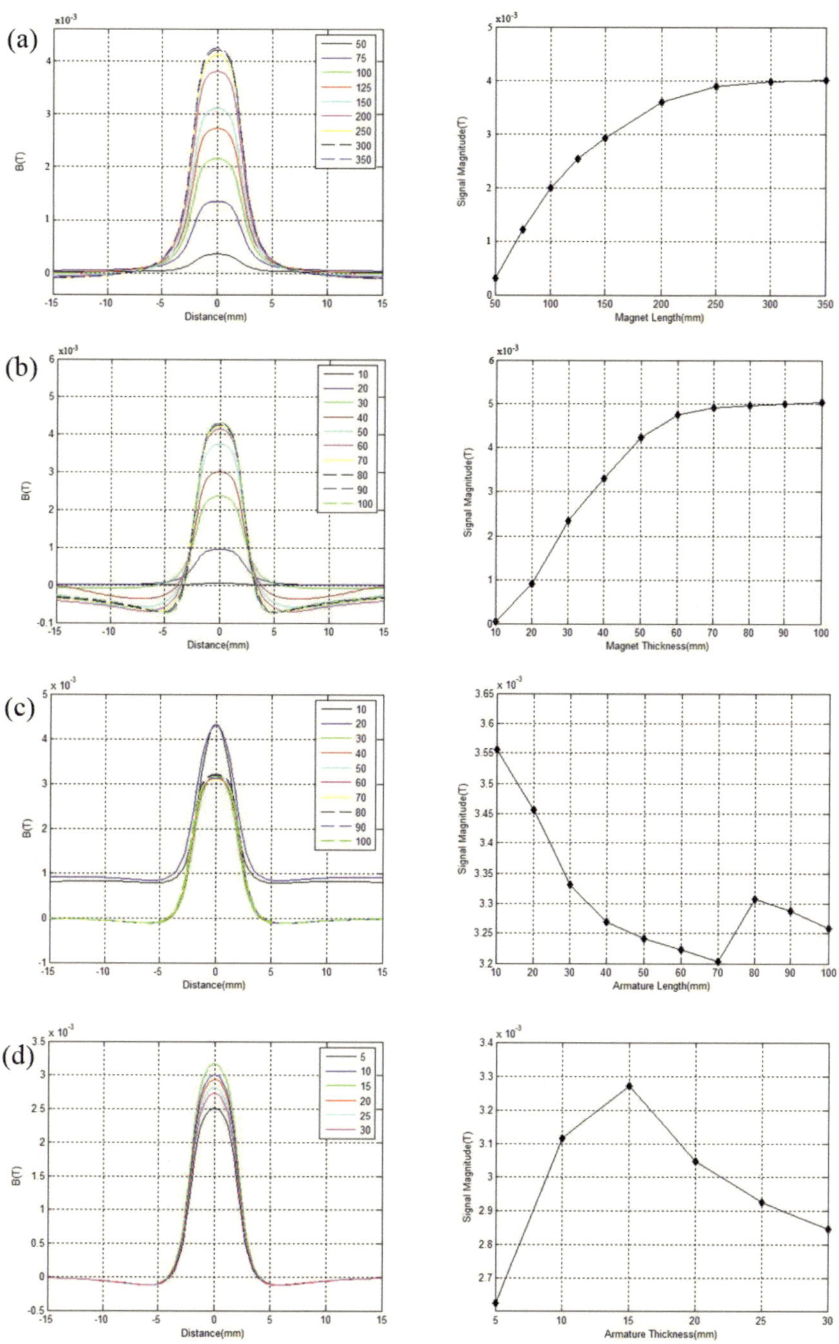

Figure 20. The MFL signals of a broken-wire defect with the changing of ring-shaped open magnetization MFL sensor sizes. (a) The MFL signals of a broken-wire defect with the changing of axial length of ring-shaped magnet. (b) The MFL signals of a broken-wire defect with the changing of radial thickness of ring-shaped magnet. (c) The MFL signals of a broken-wire defect with the changing of axial length of the ferromagnetic ring between two magnet rings. (d) The MFL signals of a broken-wire defect with the changing of radial thickness of the ferromagnetic ring between two magnet rings.

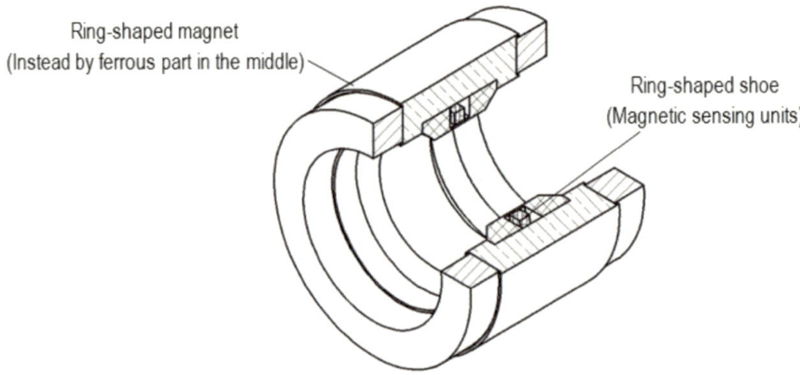

Figure 21. Open magnetization MFL sensor for steel wire rope testing.

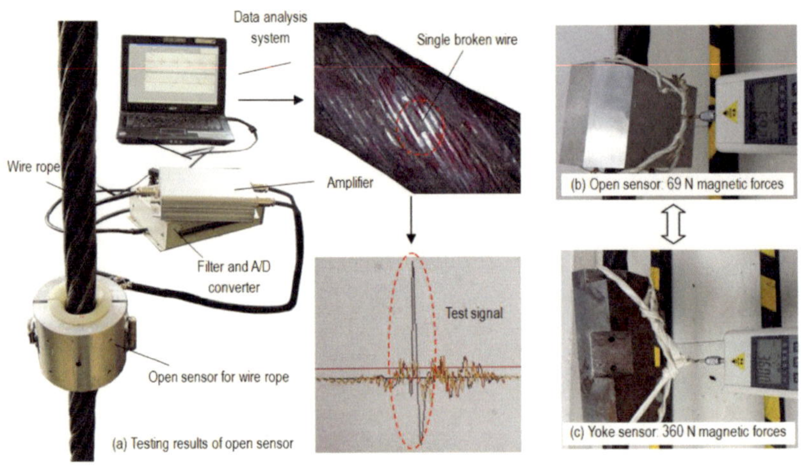

Figure 22. Experimental results of the defect detection capability and magnetic interaction forces.

sensor owns the inspection capability even for a single broken wire as the traditional yoke sensor does. Most notably, the testing for the magnetic interaction force produced by the created open sensor is only 69.2 N, while that caused by the traditional yoke one is over 360 N, which demonstrates that the new proposed open magnetization sensor features a much weaker magnetic interaction force compared with the yoke sensor and accords with the theoretically FEM analyses done before.

3. Conclusion

The life expectancy of a hoist structure largely depends on the properties of the steel wire ropes; many studies on the nondestructive evaluation of concrete are being developed.

However, many existing methodologies and techniques fail to detect steel wire rope defects, especially in sophisticated circumstances. Using tubular coils, the MFL technique has developed rapidly. However, the closed structure limits the detection scope of traditional tubular coils and prevents the testing of very long objects, such as stay cables. Moreover, a large number of studies regarding SHM focus on the electromagnetic technique, while very little literature has reported on their validity for steel wire rope detection. In short, the current electric-MFL SHM methods using tubular coils as the electro-magnetizers fail to detect steel wire ropes without a head and a tail.

Therefore, a novel electric-MFL SHM technique for concrete rebar using open electromagnetic coils as the magnetizers is presented; this open configuration is preferable for the defect detection of closed or very long structures of tested objects and has wider applications in the SHM of steel wire ropes. Based on the traditional tubular coils, a novel electromagnetic testing apparatus with an open magnetizer is proposed for theoretical analysis and comparison. Next, finite element modeling and simulation based on ANSYS are conducted, and the magnetization effect, defect detection capability, and magnetic force interaction are studied and compared. Furthermore, an optimized magnetizer structure is also presented.

To validate the reliability and validity of the simulations, two optimized magnetizers are designed and manufactured; then, experiments are conducted to determine the magnetization effect and the defect detection capability of the two magnetizing coils. The experimental results showed that the proposed open electromagnetic technique has a similar magnetization effect and an excellent defect detection capability compared to the traditional tubular coils. Finally, the consistency between the simulation and experimental results confirmed the great potential of concrete SHM using the open electromagnetic technique.

Additionally, a new sensor based on open magnetization method is characterized by weaker magnetic interaction force, and similar magnetization capability compared to the traditional yoke method is also proposed. Therefore, the novel testing sensor of the MFL testing method with these advantages is more suitable for portable defect detections of hoist wire ropes, featuring good signal-to-noise rate by reducing magnetic noises produced by rope swaying, little wear or damage and full automatic detection process. Meanwhile, the new testing sensor with preponderant characteristics further promotes the capability of addressing the challenges of portable hoist wire ropes to maintain a constant lift-off distance for scanning. The final practice applications confirm the availability and validity of the portable MFL testing apparatus for the nondestructive testing of hoist ropes under poor working conditions.

Acknowledgements

This chapter was financially supported by the National Natural Science Foundation of China (51475194), the National Key Basic Research Program of China (2014CB046706), and the Fundamental Research Funds for the Central Universities (Grant No. 2015MS015).

Author details

Shiwei Liu, Yanhua Sun* and Wenjia Ma

*Address all correspondence to: yhsun@hust.edu.cn

School of Mechanical Science and Engineering, Huazhong University of Science and Technology, Wuhan, China

References

[1] Mehrabi AB. In-service evaluation of cable-stayed bridges, overview of available methods and findings. Journal of Bridge Engineering. 2006;**11**(6):716–724

[2] Xu F, Wang X, Wu H. Inspection method of cable-stayed bridge using magnetic flux leakage detection: Principle, sensor design, and signal processing. Journal of Mechanical Science and Technology. 2012;**26**(3):661–669

[3] Jomdecha C, Prateepasen A. Design of modified electromagnetic main-flux for steel wire rope inspection. Ndt & E International. 2009;**42**(1):77–83

[4] Kwun H, Cecil MTII. Nondestructive evaluation of ferromagnetic cables and ropes using magnetostrictively induced acoustic/ultrasonic waves and magnetostrictively detected acoustic emissions: U.S. Patent 5,456,113; 1995

[5] Makar J, Desnoyers R. Magnetic field techniques for the inspection of steel under concrete cover. NDT & E International. 2001;**34**(7):445–456

[6] Zhong XY, Zhang XH. Research of on-line detection apparatus for industrial steel wire rope. Applied Mechanics and Materials. Trans Tech Publications. 2011;**48**:924–927

[7] Zawada K. Magnetic NDT of steel wire ropes. Journal of Nondestructive Testing & Ultrasonics (Germany). 1999:4(8):1–1

[8] Wait JR. Review of electromagnetic methods in nondestructive testing of wire ropes. Proceedings of the IEEE. 1979;**67**(6):892–903

[9] Sun Y, Liu S, Li R, et al. A new magnetic flux leakage sensor based on open magnetizing method and its on-line automated structural health monitoring methodology. Structural Health Monitoring. 2015;**14**(6):583–603

[10] Sun Y, Wu J, Feng B, et al. An opening electric-MFL detector for the NDT of in-service mine hoist wire. IEEE Sensors Journal. 2014;**14**(6):2042–2047

[11] Sun Y, Liu S, Deng Z, et al. Magnetic flux leakage structural health monitoring of concrete rebar using an open electromagnetic excitation technique. Structural Health Monitoring. 2017;**3**:1475921716684340

[12] Zhang W, Li S, Gao F. The modeling and simulations of the circuit element based on finite element methods. In: IEEE International Conference on Vehicular Electronics and Safety, 2005. IEEE; 2005. pp. 257–260

[13] Noreika A, Tarvydas P. Electromagnetic field modeling using edge finite elements. In: BEC 2008. 11th International Biennial Baltic on Electronics Conference, 2008. IEEE; 2008. pp. 99–102

[14] Theory Reference for ANSYS and ANSYS Workbench. ANSYS Release 11.0. SAS IP, Inc.; 2007. p. 1110

[15] Elements Reference. ANSYS Release 11.0. SAS IP, Inc.; 2007. p. 1532

[16] Hirama Y, Takahashi K, Hori S. Electromagnetic inspecting apparatus for magnetizable wire rope: U.S. Patent 4,427,940; 1984-1–24

[17] Katoh M, Masumoto N, Nishio K, et al. Modeling of the yoke-magnetization in MFL-testing by finite elements. NDT & E International. 2003;**36**(7):479–486

[18] Stupakov O, Tomáš I, Kadlecová J. Optimization of single-yoke magnetic testing by surface fields measurement. Journal of Physics D: Applied Physics. 2006;**39**(2):248

[19] Zhang Y, Ye Z, Wang C. A fast method for rectangular crack sizes reconstruction in magnetic flux leakage testing. Ndt & E International. 2009;**42**(5):369–375

[20] Radovanović ID, Rajović NM, Rajović VM, et al. Signal acquisition and processing in the magnetic defectoscopy of steel wire ropes. In: 19th Telecommunications Forum (TELFOR). IEEE; 2011. pp. 864–867

Chapter 3

Structural Damage Detection Based on Improved Multi-Particle Swarm Co-Evolution Optimization Algorithm

Shaofei Jiang and Shenglan Ma

Additional information is available at the end of the chapter

http://dx.doi.org/10.5772/intechopen.68385

Abstract

This chapter presents an improved multi-particle swarm co-evolution optimization algorithm (IMPSCO) to detect structural damage. Firstly, IMPSCO is integrated with Newmark's algorithm for damage detection and system identification, which just need few sensors. In addition, for reducing the searching parameters, a two-stage damage detection method based on modal strain energy and IMPSCO is presented. In order to validate the proposed method, a seven-story steel frame experiment in laboratory conditions is performed and a comparison is made between the proposed approach and genetic algorithm (GA). The results show that: (1) the proposed methods can not only effectively locate damage location but also accurately quantify the damage severity. Besides, it has excellent noise-tolerance and adaptability; (2) for integrating IMPSCO and Newmark's algorithm, it implements only by using any kinds of structural time-series responses and the excitation force; (3) compared with genetic algorithm, IMPSCO is more efficient and robust for damage detection with a better noise-tolerance.

Keywords: structural damage detection, improved multi-particle swarm co-evolution optimization (IMPSCO) algorithm, time-domain, frequency-domain, modal strain energy, Newmark's algorithm

1. Introduction

Structural health monitoring (SHM) and damage detection are taking an increasingly important role in civil engineering structures. As the main part of SHM programmes, damage localization and quantification strategies can provide basic information about the health of structural systems. Numerous methods have been developed and employed for damage detection based on vibration data. It can be divided in two major groups [1]. The First group are approaches in which a direct methodology is introduced for damage detection. Such an aim was achieved by performing signal processing-based procedures [2] and mathematical analysis [3]. The Second group are defining damage detection problem as an inverse problem and solving it by employing optimization procedures [4, 5].

During the last two decades, standard particle swarm optimization (PSO) and its corresponding approaches [6–9] have been applied to damage detection problems due to the virtues in global optimization. For example, Begambre and Laier [10] proposed PSOS model-based damage detection algorithm using frequency domain data to locate the damage location and quantify the damage extent. Therein, the minimization function is based on the frequency response functions of the system. Furthermore, Mohan et al. [11] also evaluated the use of frequency response function with the help of PSO for structural damage identification, which verified by beam plane frame structures. Nanda et al. [5] proposed unified particle swarm optimization (UPSO) technique for solving crack assessment problems in frame-like structures. In recent years, the multi-particle swarm co-evolution optimization (MPSCO) by integrating the collaborative theory in ecology with the principle of automatic adjustment has become a popular hotspot [12–14]. Nevertheless, such efforts cannot completely solve the problem of local optimum. Therefore, many scholars proposed many multi-stage damage detection methods, namely, firstly using other relevant method to initially locate damage location and utilizing optimization algorithm to accuracy identify damage. For instance, Seyedpoor [15] presented a two-stage damage detection method based on modal strain energy and PSO to identify damage. Vo-Duy et al. [16] presented a two-step approach based on modal strain energy method and an improved differential evolution algorithm for damage detection in laminated composite structures.

However, during practical applications, there exist some drawbacks. For example, standard PSO algorithm and improved PSO are computationally intensive because they must deal with a great number of damage variables. Therefore, an improved MPSCO algorithm called IMPSCO is proposed and applied to structural damage detection. In IMPSCO, the evolutionary theory is integrated with MPSCO as to reduce the possibility of falling into the local optimum. Furthermore, IMPSCO integrates with Newmark's algorithm in time-domain for damage detection. Because it just uses any kinds of structural time-series responses and only requires very few sensors in practical applications. In addition, this chapter also presents a two-stage method based on modal strain energy change ratio and IMPSCO algorithm for the purpose of reducing the complexity to localize and quantify the structural damage.

The organization of the chapter is as follows: Section 2 provides the basic theory of IMPSCO. Section 3 describes the damage detection strategy based on IMPSCO in time domain. Section 4 describes the two-stage damage methods based on modal strain energy and IMPSCO. Experimental study is presented in Section 5. Section 6 gives the concluding remarks.

2. Improved multi-particle swarm co-evolution optimization algorithm

With basic MPSCO, multi-subpopulations are divided into two layers. All particles from the upper-layer follow the optimum of the entire population so as to obtain a faster convergence speed, while all particles from the lower layer follow the optimum of the subpopulation to which it belongs, so as to ensure the population diversity. Although the performance of basic MPSCO is better than standard PSO in some aspects, the subpopulations in lower-layer still perform the process of standard PSO, which makes its falling into the local optimum possible.

To solve this problem, the IMPSCO algorithm is proposed and applied to locate and quantify damage of a structure in this study.

In nature, some species disappear because of environmental variation, while some new species will emerge at the same time so that the species diversity is balanced. Accordingly, the searching procedure of PSO can be regarded as a nature evolution procedure. More specifically, if a particle is recorded as the worst particle many times, it indicates that this particle is unable to meet the current requirement and needs to be eliminated. Consequently, the improved multi-particle swarm co-evolution optimization algorithm is presented, in which the particles to be deleted are replaced with the gravity position of the selected excellent particles in current entire population. After the replacement, the particles can quickly get out of the local optimum. All in all, the proposed IMPSCO algorithm is shown in **Figure 1**.

Step 1: (population initialization). Generate m subpopulations randomly, and then divide them into the upper layer with only one subpopulation and the lower layer with $m-1$ subpopulations. For each subpopulation, it contains n particles. Meanwhile, set the iteration index k to zero.

Step 2: (fitness calculation). Calculate the fitness values of each particle and save the personal best values and its corresponding particle's location as $g$best$j$ and $g$bestval$j$ ($j = 1, 2 \ldots m$), respectively. Meanwhile, record the best individual in the entire population and its corresponding fitness as $g$bestval and $g$best.

Step 3: (particles updating). Update the particles in the upper layer and the lower layer according to Eqs. (1) and (2), respectively, and the worst particle in the entire population is recorded:

$$\begin{aligned} v_i^{k+1} &= \omega v_i^k + c_1 r_1 (p_i - z_i^k) + c_2 r_2 (p_g - z_i^k) \\ z_i^{k+1} &= z_i^k + v_i^{k+1} \end{aligned} \tag{1}$$

$$\begin{aligned} v_i^{k+1} &= \omega v_i^k + c_1 r_1 (p_i - z_i^k) + c_2 r_2 (p_{lg} - z_i^k) \\ z_i^{k+1} &= z_i^k + v_i^{k+1} \end{aligned} \tag{2}$$

where i is the particle's index in the swarm; z_i and v_i represent the position and velocity of the particle, respectively; p_i represents the optimal position of the particle; p_g and p_{1g} represent the optimal position of the entire population and the subpopulation to which the particle belongs, respectively; r_1 and r_2 are the random numbers between zero and one; c_1 and c_2 are the learning factors; ω is the inertia weight.

Additionally, the maximum velocity of each particle cannot exceed V_{max} which is set to be 20% of the length of the search space.

Step 4: (optimum updating). Calculate the fitness of each updated particle and compare it with the values above. If it is preferable, then update $p$best, $p$bestval, $g$best, $g$bestval, $g$best$j$ and $g$bestval$j$, correspondingly. Let $k = k+1$.

Step 5: (worst particle replacement). Repeat Steps 3 and 4. When the particle is recorded as the worst for the predetermined times, I_w replaces it with G_g as shown in Eq. (3), and its worst record returns to zero:

$$G_g = \frac{\sum_{j=1}^{s} z_j}{s} \tag{3}$$

where s and z_j represent the number of the selected excellent particles and their position, respectively.

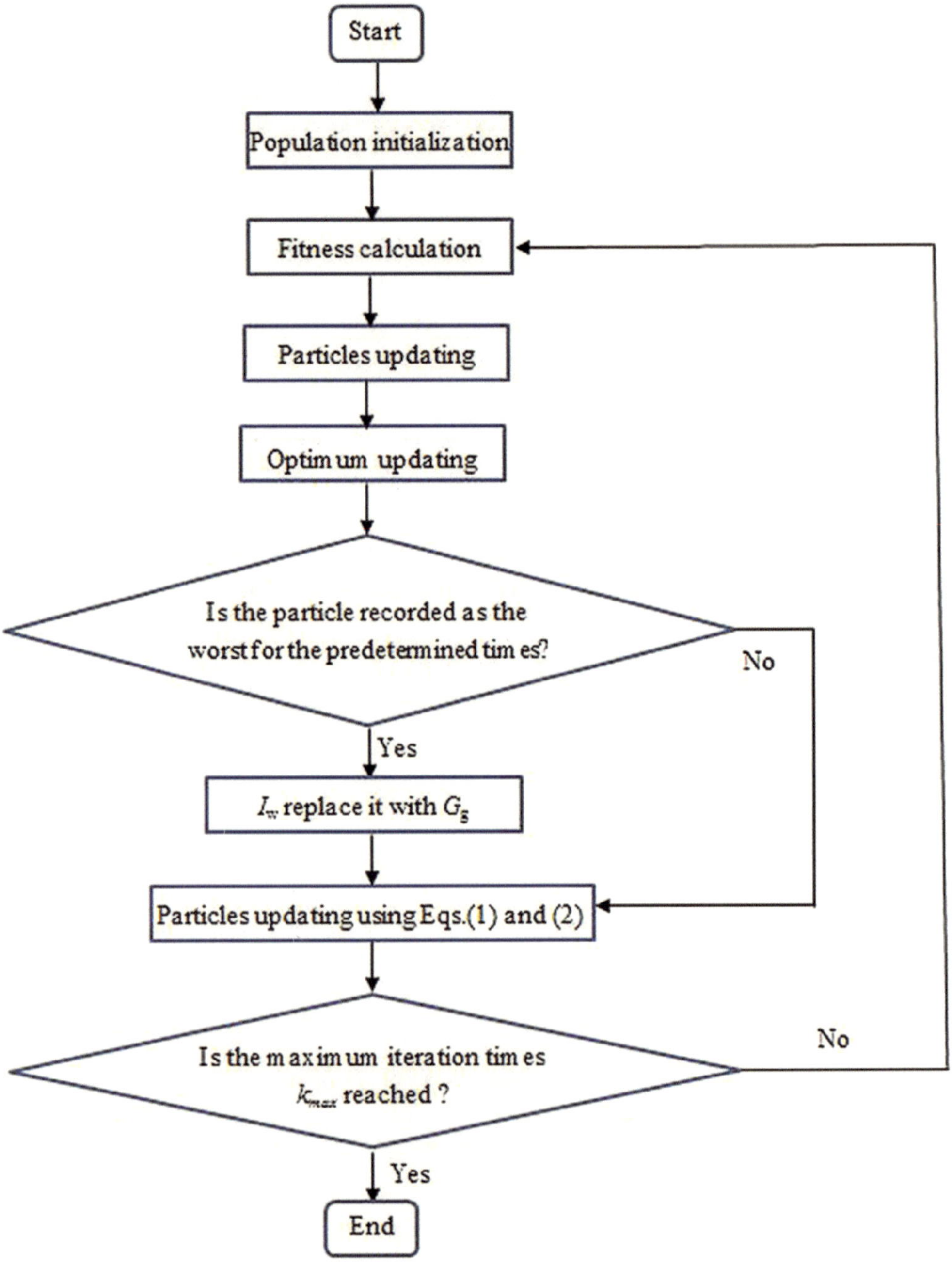

Figure 1. The flow chart of the IMPSCO algorithm.

Step 6: Go to Step 3, and repeat until the maximum iteration times $k_{\max}$ is reached.

Compared with MPSCO, IMPSCO added the process of worst particle replacement, which is able to avoid the problem of local optimum. Moreover, it can increase computing efficiency.

3. Damage detection strategy based on time-domain

On the basis of the abovementioned, this section proposes a time-domain damage detection strategy by integrating the IMPSCO and Newmark's algorithm. It contained two steps, namely, determination of damage threshold and parameter identification with IMPSCO and Newmark's algorithm [17]. For parameter identification, it involves parameters encoding, establishment of fitness function and parameters setting. The schematic diagram of the damage detection strategy based on time-domain is depicted in **Figure 2**.

3.1. Determination of damage threshold

It is difficult and significant task to determine the damage threshold because the threshold is usually used to judge damage. If the identification results are higher than the threshold, it indicates that the damage occurs; otherwise the damage is excluded because of the inevitable measurement noise. Eventually, the damage detection is performed. In this study, the damage threshold is set to 2% validated by the statistical hypothesis testing. The details can be seen in Ref. [18].

3.2. Parameters encoding

The first task is to encode parameters involving the IMPSCO. For a frame, generally, the floor stiffness is encoded directly as the parameter to be optimized in the intact state. However, the actual stiffness values cannot be obtained accurately, so we turn to focus on detecting the stiffness reduction for damage scenarios. Therefore, the floor stiffness ratio of the damaged structure to the undamaged structure, which is defined as the rigidity coefficient, is introduced and encoded to make the detection results simple and clear. In addition, with the Rayleigh

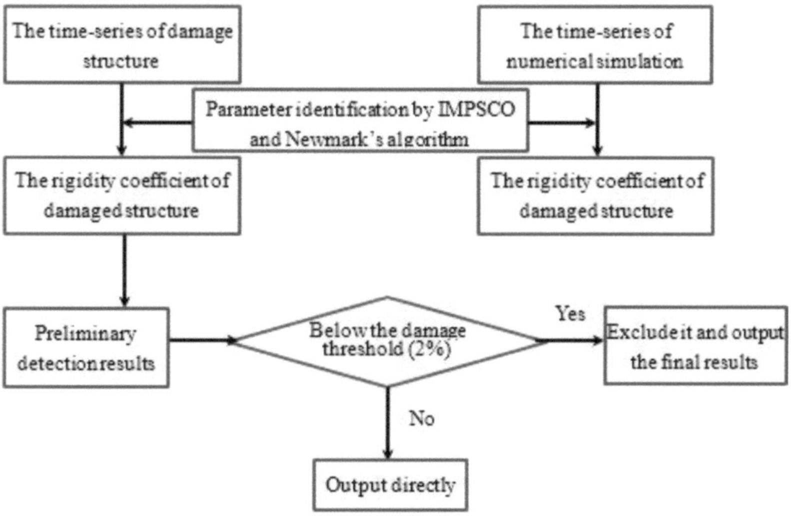

Figure 2. The schematic diagram of the damage detection strategy based on time-domain.

damping taken into consideration, the mass and stiffness damping coefficients α and β should also be encoded.

3.3. Parameters setting in IMPSCO

In this study, the parameters are set as follows: the total number of subpopulation $m = 3$; each population size $n = 10$; learning factors $c_1 = c_2 = 2$; limited times for the worst record $I_w = 10$; maximum iteration times $k_{\max} = 60$; the number of the selected excellent particles s = 6; the inertia weight w is linearly decreased from 0.9 to 0.4 before the 45th iteration and afterwards it maintains at 0.4 to enhance the local search capability.

3.4. Establishment of fitness function

The most important task is to determine the fitness function for the IMPSCO. Firstly, the structural mass matrix M, stiffness matrix K and damping matrix C are constructed with the generation of particles. And then the simulated time-series responses can be obtained by using Newmark's constant-average acceleration method. Only if the analytical responses are close to the measured ones can it be determined that the structural properties represented with the particles agree well with actual damage situations. Consequently, the fitness function can be represented with

$$F(\theta) = \frac{1}{\sum_i^N \sum_l^L (X_{i,l}^{mea} - X_{i,l}^{sim})^2} \tag{4}$$

where θ is the parameters vector; N is the number of measuring points; L is the length of response data and $X_{i,l}^{mea}$ and $X_{i,l}^{sim}$ are the measured and simulated time-series responses, respectively.

When $F(\theta)$ reaches the maximum, the values of θ are the optimal solution.

4. Two-stage damage detection strategy based on modal strain energy and IMPSCO

Just as it is mentioned above, IMPSCO integrated with Newmark's algorithm can locate damage location and quantify damage severity. However, for structural damage detection, the damage variables are always in great number. Therefore, just used IMPSCO to detection damage can lead to time consuming. For the sake of reducing the number of optimization variables in structural damage detection using IMPSCO, this chapter also presents a novel two-stage structural damage detection method by the means of combination of modal strain energy (MSE) and IMPSCO. More specifically, in the first stage, the modal strain energy is used to exclude the possible healthy variables. In the second stage, IMPSCO, regarded structural frequency as fitness function, is applied to precisely locate and quantify damage. The schematic diagram of the two-stage damage detection based on modal strain energy and IMPSCO is depicted in **Figure 3**.

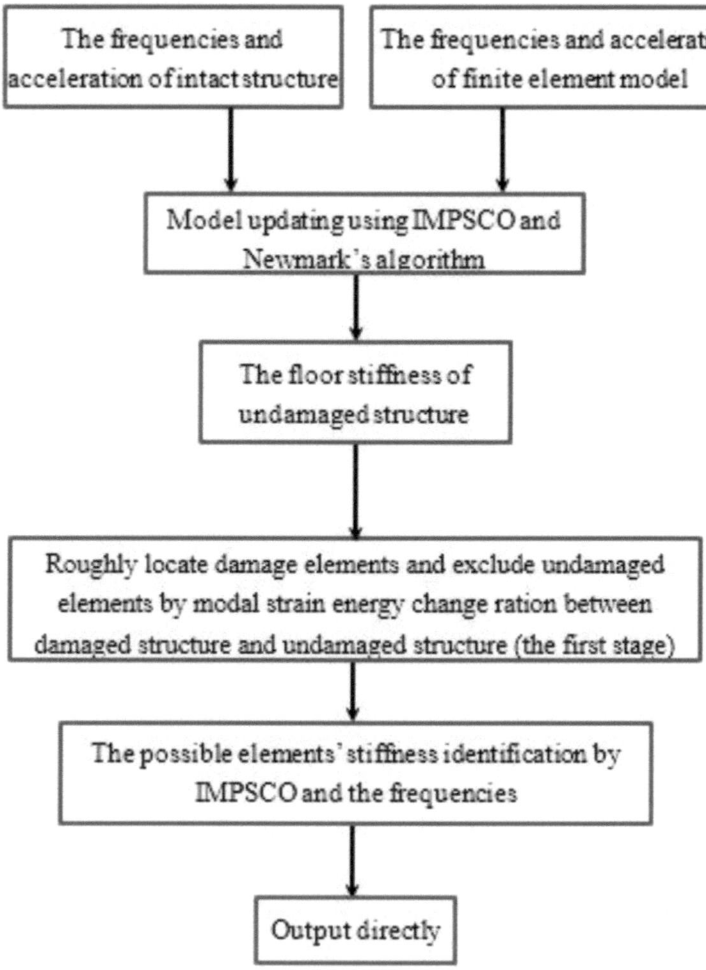

Figure 3. The schematic diagram of the two-stage damage detection based on modal strain energy and IMPSCO.

4.1. The first stage: modal strain energy

The purpose of modal analysis here is to obtain the modal parameters of the structure, such as natural frequencies and mode shapes. So, it has the mathematical form of [19]

$$(\boldsymbol{K} - \omega_i^2 \boldsymbol{M})\phi_i = 0 \qquad (\ i = 1, \ldots\ldots, n) \tag{5}$$

where $\boldsymbol{K}$ and $\boldsymbol{M}$ are the stiffness and mass matrices of the structure, respectively; ω_i and ϕ_i are the i-th circular frequency and mode shape vector of the structure, respectively. Also, n is the total degrees of freedom of the structure. Mode shapes are usually normalized with respect to the mass matrix and the relations are thus established.

The modal strain energy of the i-th element in mode j can be expressed as:

$$MSE_{ij}^{u} = \{\phi_j^u\}^T[\boldsymbol{K}_i^u]\{\phi_j^u\} \tag{6}$$

$$MSE_{ij}^{d} = \{\phi_j^d\}^T[\boldsymbol{K}_i^d]\{\phi_j^d\} \tag{7}$$

where $\boldsymbol{K}_i$ is the stiffness matrix of the i-th element of the structure and ϕ_j is the vector of corresponding nodal displacements of element i in mode j; u and d are undamaged and damaged structures, respectively.

The modal strain energy change ratio between initial structure and damage structure is used to exclude the possible undamaged elements, which can be seen as follows:

$$MSEC_{ij} = MSE_{ij}^{u} - MSE_{ij}^{d} = \{\phi_j^d\}^T[\boldsymbol{K}_i^u]\{\phi_j^d\} - \{\phi_j^u\}^T[\boldsymbol{K}_i^u]\{\phi_j^u\} \tag{8}$$

$$MSECR_{ij} = \frac{|MSEC_{ij}|}{MSE_{ij}^{u}} \tag{9}$$

$$MSECR_i = \max\left[\frac{1}{m}\sum_{j=1}^{m}\frac{MSECR_{ij}}{MSECR_{j,\max}}, 0\right] \tag{10}$$

where $MSEC_{ij}$ is the modal strain energy change of the i-th element in mode j; $MSECR_{ij}$ is the modal strain energy change ratio of the i-th element in mode j; $MSECR_i$ is the modal strain energy change ratio of the i-th element; $MSECR_{j,\max}$ is the maximum of the absolute value of the $MSECR_{ij}$ in mode j.

It should be noted that the damaged elements are still unknown for a damaged structure in this stage; therefore, the element stiffness matrix of the intact structure is used for estimating the parameter $MSECR_{ij}$ in this chapter. So a damage threshold calculation method as shown in Ref. [19] is utilized to determine the modal strain energy change ratio threshold. In other words, the $MSECR_i$ will be less than the threshold for an undamaged variable, but greater than the threshold for the possible damaged variable.

4.2. The second stage: identifying the damage accurately

In this stage, IMSPCO presented in Section 2 is used to identify the damaged elements accurately as well as to determine the extent of damage. In addition, the procedure and parameters of IMSPCO are the same as Section 3. But the minimization objective function is using the error of frequencies between measured and analysed. The details can be seen as follows:

$$\boldsymbol{F} = \frac{1}{\sum_{j=1}^{r}(\omega_j^* - \omega_j)^2} \tag{11}$$

where ω_j^* denotes the measured circular frequency of the j-th mode, which is calculated from vibration testing; ω_j is the analytical circular frequency of the j-th mode, which can be calculated as follows:

$$\omega_j^2 = \frac{\{\phi_j\}^T[\boldsymbol{K}^u - \Delta\boldsymbol{K}]\{\phi_j\}}{\{\phi_j\}^T[\boldsymbol{M}]\{\phi_j\}} \tag{12}$$

Where ϕ_j is the experimental mode shape vector of the j-th mode; K^u and M are the stiffness and mass matrices of the intact structure, respectively; ΔK is the stiffness matrix changes due to structural damage, which can be represented with:

$$\Delta\boldsymbol{K}_i = b_i \times \boldsymbol{K}_i^u \tag{13}$$

in which, b_i is the damage extent of i-th element of the structure, which can be represented with:

$$b_i = \frac{K_i^u - K_i^d}{K_i^u} \times 100\% \tag{14}$$

5. Experimental study

5.1. Steel frame

A seven-floor steel frame of 1.4125 m in height is constructed and tested in the laboratory [20]. The model is designed with flexible columns and relatively rigid beams to simulate a shear building, as shown in **Figure 4**. The cross-section properties of the structural members are listed in **Table 1**. The mass of the structure is concentrated at the floors, and the structural model is regarded as a lumped mass model. Therefore, the steel frame can be simplified as a 7 degree of freedom spring-mass system with the masses $M_1 \ldots\ldots M_6 = 3.78$ kg and $M_7 = 3.3$ kg.

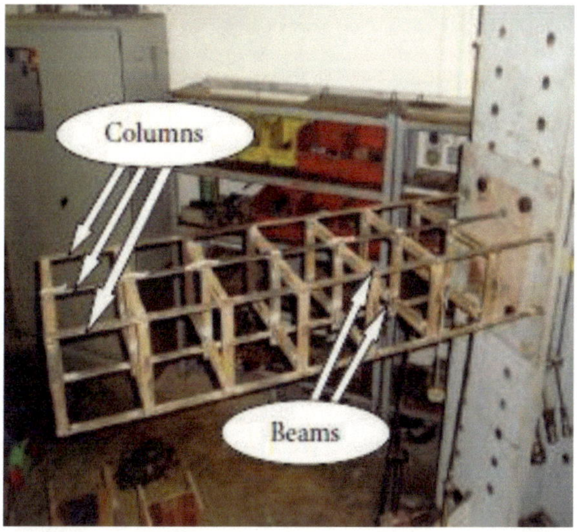

Figure 4. Seven-floor steel frame model.

Parameter	Beams	Columns
Sectional dimension	$25 \times 25 \times 3$ *mm* (SHS)	25×4.6 *mm*
Cross-sectional area A (m^2)	264×10^{-6}	115×10^{-6}
Cross-sectional moment of inertia I (m^4)	2.17×10^{-8}	2.03×10^{-10}
Elasticity modulus E (Pa)	206×10^9	206×10^9
Density ρ (kg/m^3)	7850	7850

Table 1. Structural properties.

5.2. Dynamic test

1. Damage scenarios: As the structure is constructed with six columns per floor, damage is simulated by cutting the centre column partially or completely to keep the symmetry of the model. Small damage is produced by four partial cuts near the top and bottom of the column, whereas large damage is simulated by a complete cut at the mid-height of the column, as indicated in **Figure 5**. The expected reduction of floor stiffness due to the small damage is estimated by the software of ABAQUS. The finite element models of both the undamaged and damaged columns are established, and the displacements under the same nodal loads are calculated and compared to determine the change in column stiffness. As there are six columns in each floor, the small damage results in a reduction in the floor stiffness with 4.1%. As for the case of large damage, the column damage is 100% resulting in a reduction in floor stiffness of $1/6 \approx 16.7\%$. More details can be seen in the literature [20]. The two damage scenarios are shown in **Table 2**.

2. Experimental data acquisition: **Figure 6** illustrates the dynamic testing and data acquisition system. For ease of setup, the model is mounted horizontally and excited by vibration exciter vertically at the seventh floor. The force generated is measured by an Integrated Circuit Piezoelectric (ICP) force sensor (model PCB-208C02). Meanwhile, the structural responses are measured using the ICP accelerometers mounted at each floor. The data are recorded at a sampling frequency of 5,000 Hz.

5.3. Damage detection using IMPSCO combined with Newmark's algorithm

5.3.1. Identification of undamaged structure (Case 1)

Firstly, 500 data points of the acceleration responses at each floor and the corresponding force are extracted by the installed sensors when the structure is intact. Then follow the steps presented in **Figure 1**. Consequently, the floor stiffness is identified for the intact structure.

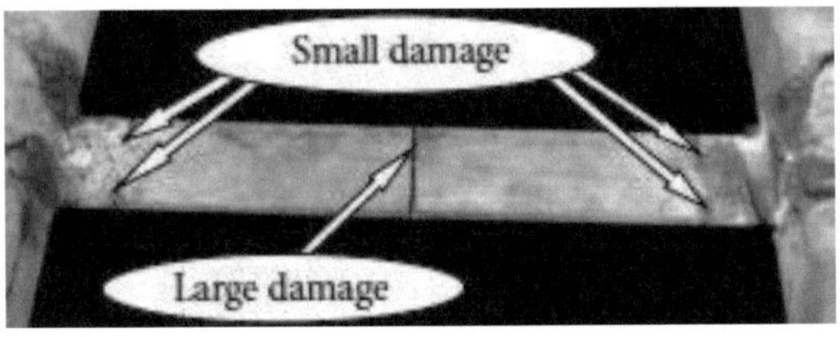

Figure 5. Damage applied to the column.

Case no.	Small damage	Large damage
Case 0	–	–
Case 2	–	Story 4
Case 3	Story 6	Story 4

Table 2. Damage scenarios.

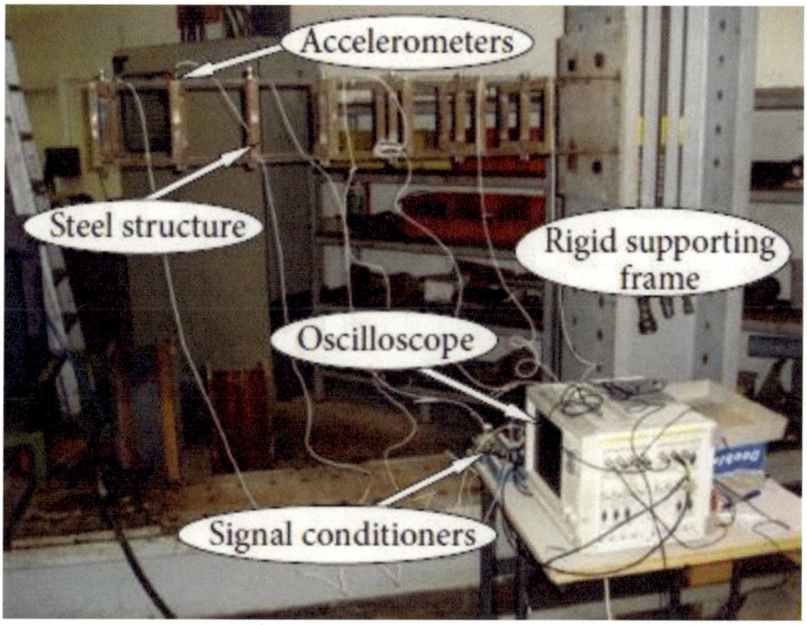

Figure 6. Test setup.

Identification of undamaged structure: For the intact structural model, each floor stiffness and damping coefficient are encoded as $\boldsymbol{\theta} = [K1, K2, \ldots K7, \alpha, \beta]$. The search range of Ki ($i = 1, 2, \ldots, 7$) is within the interval of [0–800 kN/m], and the search range of α and β is within the intervals of [0–4] s−1 and [0–0.001]s, respectively. By following the steps presented in Section 3, the floor stiffness of the intact structure can be identified.

Because the IMPSCO algorithm is a probabilistic optimization algorithm, the evolutionary process is always accompanied with randomness. Consequently, it is difficult to judge whether the detection result is correct from a single test.

The detection process is repeated for 15 times using the same method in order to improve the identification precision. The average values for the 15 times are regarded as the final results, as shown in **Table 3**.

It can be seen from **Table 3** that the identification result of k_1 is significantly lower, and this is due to the less rigid connection at the base of the structure or the modelling errors of the initial numerical model. In addition, the identification results of α and β are 0.75102223 and 0.00000286, respectively.

Floor stiffness	K1	K2	K3	K4	K5	K6	K7
Identification (kN/m)	300	626	553	566	547	574	519

Table 3. Identification results of floor stiffness in intact state.

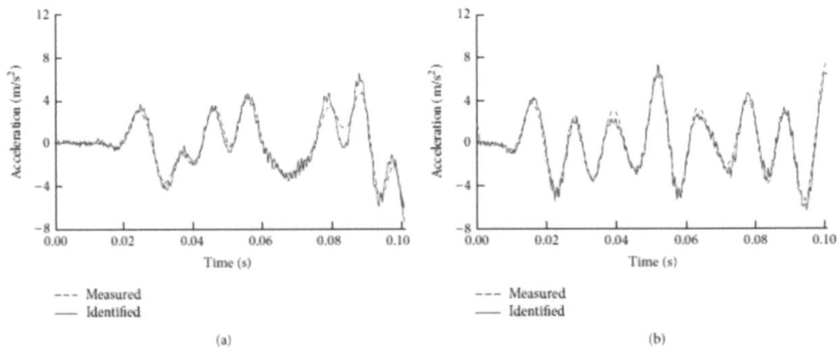

Figure 7. The comparison results of acceleration response. (a) The second floor; (b) the fifth floor.

Then the acceleration responses can be calculated by the identified results according to Newmark's algorithm, which is plotted along with the measured values in **Figure 7**. For the simplicity, only the comparison results of the second and fifth floors are given as shown in **Figure 7**. It indicates that the identified and measured acceleration responses are in good agreement, which validates the reasonability of applying the identified floor stiffness of the undamaged structure to the detection of the rigidity coefficients under the different damage scenarios.

5.3.2. Identification of damaged structure

The identification procedure is similar to that of intact structure in Section 5.3.1, except that the parameters encoded are replaced with $\theta = [R_1, R_2, \ldots, R_7]$, in which $R_i = 1 - b_i \ (i = 1, 2, \ldots, 7)$ represents the rigidity coefficient and belongs to [0, 1].The final identification results are listed in **Table 4**.

It is seen from **Table 4** that the maximum relative error of identification results is 1.4%, which occurs in the second floor for the Case 2. As the rigidity coefficient introduced in the parameter encoding reflects the structural stiffness reduction, the effect on the damage identification results owing to the modelling error of the initial numerical model can be reduced to a great extent. This implies that the proposed structural damage detection strategy is effective and reliable.

5.3.3. Comparison and discussion

A comparison is made among the theoretical damage extent, identification results by genetic algorithm (GA) and IMPSCO, and the results are shown in **Figure 8**.

On the basis of damage threshold determined in Ref. [16], the floor whose damage extent is lower than 2% is intact. It is found from **Figure 8** that damage locations can be correctly localized

Damage scenarios	R1	R2	R3	R4	R5	R6	R7
Case 2							
Identification	0.993	0.986	0.993	0.839	0.999	0.993	0.998
Theoretical values	1.000	1.000	1.000	0.833	1.000	1.000	1.000
Relative error (%)	0.7	1.4	0.7	0.7	0.1	0.7	0.2
Case 3							
Identification	0.994	0.994	0.995	0.843	0.992	0.966	1.000
Theoretical values	1.000	1.000	1.000	0.833	1.000	0.959	1.000
Relative error (%)	0.6	0.6	0.5	1.2	0.8	0.7	0.0

Table 4. Identification results of all damage scenarios.

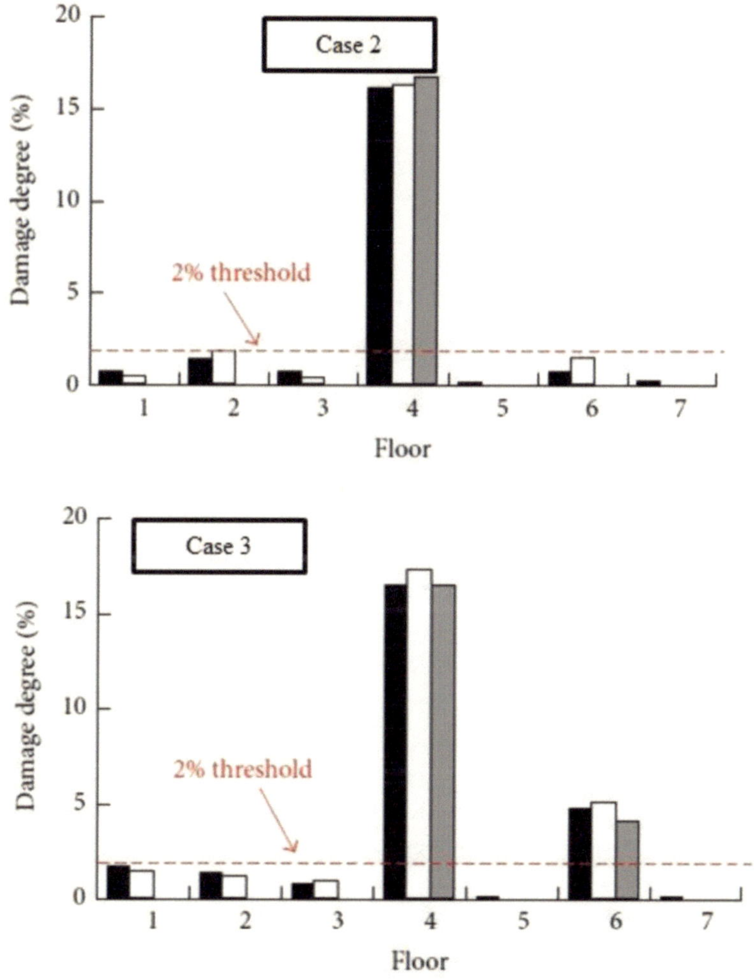

Figure 8. Identification results of all damage scenarios.

using IMPSCO and GA for all damage scenarios. In addition, the detection of damage extent is also accurate and reliable using the two optimization methods. Furthermore, IMPSCO is more efficient and accurate than GA in general. Although GA is capable of detecting damage, it costs more running time than IMPSCO.

All in all, the proposed damage detection strategy can not only localize the damage correctly but also quantify damage precisely.

5.4. Damage detection using model strain energy and IMPSCO

5.4.1. Identification of undamaged structure

Because the advantage of the two-stage damage detection method based on model strain energy and IMPSCO is that it can filter the undamaged elements. For modal updating, all elements are the possible variables. Therefore, in this section, we just used the result in Section 5.3.1 as intact structure.

5.4.2. Identification of damaged structure

5.4.2.1. Case 2

In the first stage, the modal strain energy change ratio of 7 elements is analysed, and then the index *MSECR* is evaluated. Specifically, the circular frequencies and mode shapes in intact and damaged cases are obtained by Fast Fourier Transform (FFT), and then the *MSECR* of the first four modes are calculated by Eq. (10) and depicted as in **Figure 9(a)**. Therein, the first four modes are used to calculate modal strain energy. The line of the *MSECR* threshold also plotted in **Figure 9(a)**. The value is set to 0.14 [21]. We can see from **Figure 9(a)** that the possible damaged elements are both elements 4 and 5 for Case 2.

In the second stage, IMPSCO is then employed to solve the reduced damage elements optimization detection problem to precisely locate damage location and quantify damage severity. Therein, the searching elements in IMPSCO algorithm are the possible damaged elements located in the first stage, namely elements 4 and 5. **Figure 9(b)** shows the final damage identification result. It is observed from **Figure 9(b)** that the identified damage is 0.171 for element 4 and the induced damage is 0.167 for element 4. The relative error between identified and induced damage is 2.4%. This indicates that the proposed IMPSCO algorithm can detect damage correctly.

5.4.2.2. Case 3

Case 3 is a multi-damage pattern that involves large damage scenarios in element 4 and small damage scenarios in element 6. As detection of the process for Case 2, the first four modes of the *MSECR* values are presented in **Figure 10(a)**, from which it can be seen that the values of *MSECR* for elements 4, 5 and 6 exceed the *MSECR* threshold. Therefore, it can be concluded that the possible damaged elements are elements 4, 5 and 6.

After roughly determining damage locations, IMPSCO is then utilized to accurately identify damage. The identification results for Case 3 are depicted in **Figure 10(b)**. The identified

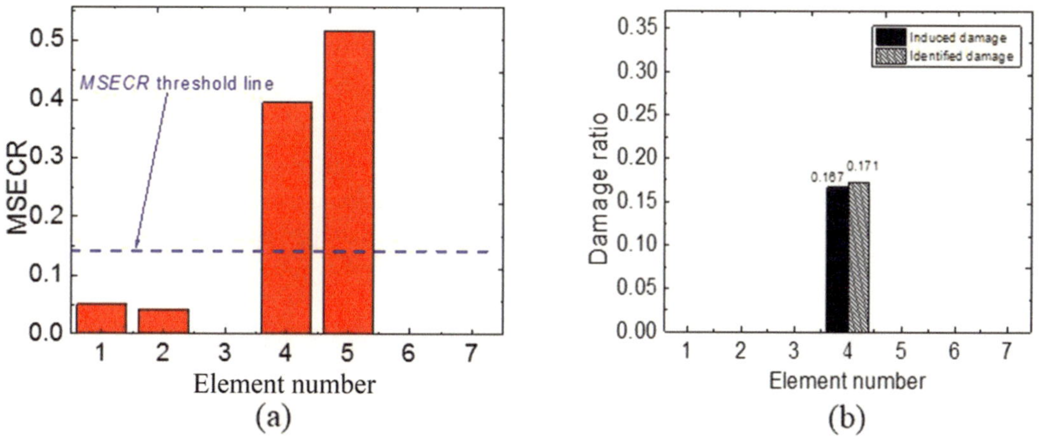

Figure 9. Identification results for Case 2. (a) Value of *MSECR*; (b) ientification results.

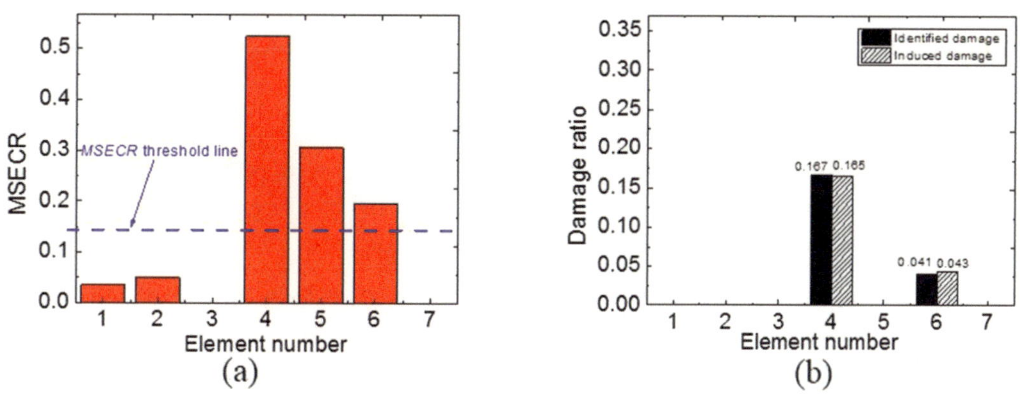

Figure 10. Identification results for Case 3. (a) Value of *MSECR*; (b) ientification results.

damaged severity is close to that of the induced damage, and the relative errors are 1.2 and 4.2% for element 4 and 6, respectively. This indicates that the proposed method is effective in precisely locating and quantifying damage.

5.4.3. Discussion

It can be seen from **Figures 9** and **10** that element 5 (Story 5) experienced an incorrect location in the first stage. There are two possible reasons for the misdetection. The first one is that the element 5 is adjacent to damaged elements (element 4 for Case 2 as well as elements 4 and 6 for Case 3) and thus its values of modal strain energy may be higher than that of other undamaged elements. In addition, the modal strain energy threshold is ensured by sampling survey, which is performed by numerical simulation. The precision of the threshold is also related to the sample size. The second reason is that subtle differences exist in the modal parameters between the finite model and experimental one for the intact structure. All in all, the efficiency of the proposed two-stage damage detection method depends on the precision of the finite element model and sample size in ensuring modal strain energy threshold.

From **Figures 9(b)** and **10(b)**, the damage severity is 0 for element 5. Specifically, this is the damage detection results by use of IMPSCO in the second stage. As a consequence, it can be concluded that IMPSCO can accurately locate and quantify damage. In summary, the proposed two-stage damage detection method is applicable to identify damage location and its severity.

6. Concluding remarks

This chapter proposes an improved multi-particle swarm co-evolution optimization (IMPSCO), and combined it with Newmark's algorithm. In addition, this chapter also presents a novel two-stage damage detection method based on modal strain energy and IMPSCO to locate damage location and quantify damage extent. The following conclusions can be made:

1. The performance comparison has proved that IMPSCO algorithm is more precise and faster than standard MPSCO in damage detection. Moreover, it has better noise-tolerance and robustness than standard MPSCO.
2. The proposed IMPSCO algorithm integrated with Newmark's algorithm is applicable and effective for detecting and quantifying damages using the noise polluted measured data. It is noted that the proposed strategy is implemented only by using any kinds of structural time-series responses and the excitation force.
3. The combination of modal strain energy and IMPSCO algorithm can provide an efficient and fast tool for properly detecting damage in simulation test and experimental example.
4. Compared with genetic algorithm, IMPSCO is more efficient and robust for damage detection with a better noise-tolerance.

Author details

Shaofei Jiang* and Shenglan Ma

*Address all correspondence to: cejsf@fzu.edu.cn

College of Civil Engineering, Fuzhou University, Fuzhou, China

References

[1] Hosseinzadeh AZ, Amiri GG, Razzaghi SAS, Koo KY, Sung SH. Structural damage detection using sparse sensors installation by optimization procedure based on the modal flexibility matrix. Journal of Sound and Vibration. 2016;**381**:65–82

[2] Ge M, Lui EM. Structural damage identification using system dynamics properties. Computer and structures. 2005;**83**(27):2185–2196

[3] Bagheri A, Ghodrati Amiri G, Seyed Razzaghi SA. Vibration-based damage identification of plate structures via curvelet transform. Journal of Sound and Vibration. 2009;**327**:593–603

[4] Chou JH, Ghaboussi J. Genetic algorithm in structural damage detection. Computers and Structures. 2001;**79**(14):1335–1353

[5] Nanda B, Maiti DK. Crack assessment in frame structures using modal data and unified particle swarm optimization technique. Advances in Structural Engineering. 2014;**17**(5):747–766

[6] Kennedy J, Eberhart R. Particle swarm optimization. In: IEEE Service Center, editor. In Proceedings of the IEEE International Conference on Neural Networks; Piscataway. USA: 1995. pp. 1942–1948

[7] Chen DZhao C. Particle swarm optimization with adaptive population size and its application. Applied Soft Computing. 2009;**9**(1):39–48

[8] Chen MR, Li X, Zhang X, Lu YZ. A novel particle swarm optimizer hybridized with extremal optimization. Applied Soft Computing. 2010;**10**(2):367–373

[9] Nickabadi A, Ebadzadeh MM, Safabakhsh R. A novel particle swarm optimization algorithm with adaptive inertia weight. Applied Soft Computing. 2011;**11**(4):3658–3670

[10] Begambre O, Laier JE. A hybrid particle swarm optimization–simplex algorithm (PSOS) for structural damage identification. Advances in Engineering Software. 2009;**40**(9): 883–891

[11] Mohan SC, Maiti DK, Maity D. Structural damage assessment using FRF employing particle swarm optimization. Applied Mathematics and Computation. 2013;**219**(20):10387–10400

[12] Li AG. Particle swarms cooperative optimizer. Journal of Fudan University. 2004;**43**(5):923–925

[13] Xu K, Liu D. Algorithm of multi-PSO co-evolution based on GA and PSO. Computer Engineering and Applications. 2009;**45**(3):51–54

[14] Man CT, Sheng CM. An improved algorithm based on cooperative particle swarm optimization. Journal of Harbin University of Science and Technology. 2010;**15**(6):51–53

[15] Seyedpoor SM. A two stage method for structural damage detection using a modal strain energy based index and particle swarm optimization. International Journal of Non-Linear Mechanics. 2012;**47**(1):1–8

[16] Vo-Duy T, Ho-Huu V, Dang-Trung H, Nguyen-Thoi T. A two-step approach for damage detection in laminated composite structures using modal strain energy method and an improved differential evolution algorithm. Composite Structures. 2016;**147**(1):42–53

[17] Newmark NM. A method of computation for structural dynamics. Journal of the Engineering Mechanics Division. 1959;**85**(1):67–94

[18] Jiang SF, Wu SY, Dong LQ. A time-domain structural damage detection method based on improved multiparticle swarm coevolution optimization algorithm. Mathematical Problems in Engineering. 2014;**2014**(2):77–85

[19] Shi ZY, Law SS, Zhang LM. Structural damage localization from modal strain energy change. Journal of Sound and Vibration. 1998;**218**(5):825–844

[20] Koh CG, Perry MJ. Structural Identification and Damage Detection using Genetic Algorithms. London, UK: Structures and Infrastructures Series; 2010

[21] Ma SL, Jiang SF, Weng LQ. Two-stage damage identification based on modal strain energy and revised particle swarm optimization. International J of Structural Stability and Dynamics. 2014;**14**(5):1440005

Chapter 4

Operational Modal Analysis of Super Tall Buildings by a Bayesian Approach

Feng-Liang Zhang and Yan-Chun Ni

Additional information is available at the end of the chapter

http://dx.doi.org/10.5772/intechopen.68397

Abstract

Structural health monitoring (SHM) has attracted increasing attention in the past few decades. It aims at monitoring the existing structures based on data acquired by different sensor networks. Modal identification is usually the first step in SHM, and it aims at identifying the modal parameters mainly including natural frequency, damping ratio and mode shape. Three different field tests can be used to collect data for modal identification, among which, ambient vibration test is the most convenient and economical one since it does not require to measure input information. This chapter will focus on the operational modal analysis (OMA), i.e. ambient modal identification of four super tall buildings by a Bayesian approach. A fast frequency domain Bayesian fast fourier transform (FFT) approach will be introduced for OMA. In addition to the most probable value (MPV) of modal parameters, the associated posterior uncertainty will be also investigated analytically. The field tests will be presented and the difficulties encountered will be discussed. Some basic dynamic characteristics will be investigated and discussed. The studies will provide baseline properties of these super tall buildings and provide a reference for future condition assessments.

Keywords: structural health monitoring, operational modal analysis, ambient vibration test, super tall buildings, posterior uncertainty, Bayesian method

1. Introduction

Structural health monitoring has attracted increasing attention in the past few decades [1–3]. In the past decade, much more super tall buildings have been constructed and most of them were instrumented structural health monitoring (SHM) systems, for example, Canton Tower with the height of 610 m, on which more than 700 sensors were installed to form a sophisticated long-term SHM systems [3, 4]; Burj Khalifa, the tallest building in the world with the height of

828 m, on which an integrated real-time SHM system was instrumented [5]; Shanghai Tower with the height of 632 m and so on. Some super tall buildings are still under construction, such as Pingan International Financial Center with a height of 660 m, Shenzhen, China; Wuhan Greenland Center with a height of 636 m, Wuhan, China; Kingdom Tower with the height of 1007 m, Jeddah, Saudi Arabia. These systems play an important role in monitoring the structural conditions and making an assessment when extreme events happen. For some super tall buildings, although their heights are also more than 300 m, no long-term SHM systems were installed. Some short-term monitoring was also carried out [6, 7], for example, Building A with the height of 310 m, Hong Kong; Building B with the height of 320 m, Hong Kong; International Finance Centre with the height of 416 m, Hong Kong; etc. When some typhoons came, vibration data were collected to investigate the dynamic characteristics of these buildings.

For these super tall buildings, the monitoring of acceleration response plays an important role since the dynamic characteristics mainly including natural frequencies, damping ratios and modes shapes can be determined. They are the typical characteristics of the structures and will remain unchanged if there is no significant damage and can be used for model updating, damage detection and SHM [8–10]. The natural frequencies can help to find potential vibration problem, e.g. resonance, and then some measures can be designed to alleviate them. The damping ratio can affect the vibration level and energy dissipation capability of the structure. In the structure designing, this quantity is usually set to be constant, e.g., 5%. However, in reality, it can be affected by many factors, for example, vibration amplitude. Therefore, knowing damping ratio can better assess the structural response and provides a reference in the structural design. The mode shapes can reflect the stiffness and mass distribution and the boundary conditions of objective structures, whose change across extreme events can be used for damage detection and SHM.

To obtain the modal parameters, modal identification needs to be carried out using structural response. Acceleration and velocity data are commonly used for modal identification. There are three main ways to collect structural responses, which are free vibration test, forced vibration test and ambient vibration test. The first one is to collect structural response when the structure is under free vibration dominantly. However, for the structures in civil engineering, it is usually difficult for them to vibrate freely. The second one is to collect structural response when given some known excitation during the test. This method can improve the vibration level and provide a high signal to noise ratio for the structural response. However, to produce forced vibration, special equipment such as a high payload shaker is required. This makes the test expensive and some unexpected damage may be caused if the equipment cannot be well controlled. The third one, compared with the former two tests, is more convenient and economical since no additional excitations are required. The excitations are mainly from the ambient loading, e.g. wind, traffic loading, human walking, environmental noise, and so on. For this test, it requires high quality sensors with a lower noise level to improve the signal to noise ratio. If the noise level is too high, the modal parameters identified may have higher uncertainty.

Ambient modal identification, also called operational modal analysis (OMA), can be used to obtain the modal parameters using ambient vibration data [11]. The excitation is usually assumed to be stochastic stationary. Many methods have been developed for OMA, for example, stochastic subspace identification (SSI) [12], peak-picking (PP), and frequency domain decomposition (FDD) [13]. In addition to these non-Bayesian methods, a series of Bayesian methods have been developed [14–15], for example, Bayesian spectral density approach, Bayesian time domain method, and Bayesian fast fourier transform (FFT) approach. Recently, a fast Bayesian FFT method has been developed based on Bayesian FFT approach [16–20]. It has been extended to forced vibration data [21], free vibration data [22–23], and so on. This method views modal identification as an inference problem where the probability is taken as a measure for the relative plausibility of outcomes given a model of the structure and measured data. Both the most probable value (MPV) and the associated posterior uncertainty can be determined, making it possible to assess the accuracy of the identified modal parameters.

Based on the technique of OMA, many super tall buildings have been studied, for example, Canton Tower, super tall buildings in Hong Kong, and so on. This chapter presents the work on OMA of four different super tall buildings, including two super tall buildings in Hong Kong, Canton Tower and Shanghai Tower. It is organized as follows. In Section 2, the fast Bayesian FFT method will be introduced and it will be used for the latter study in the MPV and the associated posterior uncertainty. In Section 3, the OMA of two super tall buildings in Hong Kong will be presented. The data were collected during different strong wind events. The amplitude-dependence behaviour of modal parameters was investigated. In Section 4, Canton Tower was studied using the data collected during one day's measurement by the SHM system installed. In Section 5, Shanghai Tower, the second tallest building in the world until now, was investigated by ambient vibration tests in different stages. Finally, the summary is given in Section 6.

2. Fast Bayesian FFT method

2.1. Bayesian method for single setup

A fast Bayesian FFT approach is employed to identify the modal properties of the instrumented building using the recorded ambient vibration data in a single setup. The theory is briefly described here. The reader is referred to Ref. [14] for the original formulation and to Refs. [11, 16–18] for a recently developed fast algorithm that allows practical implementation.

The measured acceleration data are assumed to consist of the structural ambient vibration signal and prediction error

$$\ddot{\mathbf{y}}_j = \ddot{\mathbf{x}}_j + \mathbf{e}_j \tag{1}$$

where $\ddot{\mathbf{x}}_j \in R^n$ and $\mathbf{e}_j \in R^n$ (j = 1,2,..., N), respectively, denote the theoretical structural acceleration response and the prediction error, n measured degrees of freedom (DOFs), N the sampling points number. The FFT of the collected data $\ddot{\mathbf{y}}_j$ is given by

$$\mathcal{F}_k = \sqrt{\frac{2\Delta t}{N}} \sum_{j=1}^{N} \ddot{\mathbf{y}}_j \exp\left[-2\pi \mathbf{i} \frac{(k-1)(j-1)}{N}\right] \tag{2}$$

in which $\mathbf{i}^2 = -1$, Δt the sampling interval, $k = 1, \ldots, N_q$ with $N_q = \text{int}[N/2] + 1$, int[.] the integer part.

Let $\boldsymbol{\theta}$ denote the targeted modal parameters, including f_i, ζ_i, $\boldsymbol{\varphi}_i$, $i = 1, \ldots, m$ (m is the number of modes), where f_i and ζ_i denote, respectively, the natural frequency and damping ratio of the i-th mode, and $\boldsymbol{\varphi}_i \in R^n$ is the i-th mode shape vector; $\mathbf{S} \in R^{m\times m}$, S_e, the (symmetric) power spectral density (PSD) of modal forces and the PSD of prediction error, respectively.

Let $\mathbf{Z}_k = [\mathrm{Re}\mathcal{F}_k; \mathrm{Im}\mathcal{F}_k] \in R^{2n}$ denote the real and imaginary parts of the FFT data at f_k, where f_k is the FFT frequency abscissa. When performing the modal identification, the FFT data in a selected frequency band are used and they are denoted by $\{\mathbf{Z}_k\}$. Based on Bayes' theorem, the posterior probability density function (PDF) of $\boldsymbol{\theta}$ can be expressed as

$$p(\boldsymbol{\theta}|\{\mathbf{Z}_k\}) \propto p(\boldsymbol{\theta}) p(\{\mathbf{Z}_k\}|\boldsymbol{\theta}) \tag{3}$$

in which $p(\boldsymbol{\theta})$ denotes the prior PDF. The prior information is assumed to be uniform, and so the posterior PDF can be taken to be proportional to the 'likelihood function' $p(\{\mathbf{Z}_k\}|\boldsymbol{\theta})$ directly. The 'most probable value' (MPV) of $\boldsymbol{\theta}$ can be obtained by maximizing $p(\{\mathbf{Z}_k\}|\boldsymbol{\theta})$.

When N is large and Δt is small, the FFT can be shown to be asymptotically independent at different frequencies and follows a Gaussian distribution [14]. Therefore, $p(\{\mathbf{Z}_k\}|\boldsymbol{\theta})$ can be expressed as

$$p(\{\mathbf{Z}_k\}|\boldsymbol{\theta}) = \prod_k \frac{1}{(2\pi)^n (\det \mathbf{C}_k)^{1/2}} \exp\left[-\frac{1}{2}\mathbf{Z}_k{}^T \mathbf{C}_k{}^{-1} \mathbf{Z}_k\right] \tag{4}$$

in which det(.) denotes the determinant, $\mathbf{C}_k$ the covariance matrix of $\mathbf{Z}_k$ given by

$$\mathbf{C}_k = \frac{1}{2}\begin{bmatrix} \boldsymbol{\Phi} & \\ & \boldsymbol{\Phi} \end{bmatrix} \begin{bmatrix} \mathrm{Re}\mathbf{H}_k & -\mathrm{Im}\mathbf{H}_k \\ \mathrm{Im}\mathbf{H}_k & \mathrm{Re}\mathbf{H}_k \end{bmatrix} \begin{bmatrix} \boldsymbol{\Phi}^T & \\ & \boldsymbol{\Phi}^T \end{bmatrix} + \frac{S_e}{2}\mathbf{I}_{2n} \tag{5}$$

and $\mathbf{I}_{2n} \in R^{2n}$ denotes the identity matrix, $\boldsymbol{\Phi} = [\boldsymbol{\varphi}_1, \boldsymbol{\varphi}_2, \ldots, \boldsymbol{\varphi}_m] \in R^{n\times m}$ the mode shape matrix, $\mathbf{H}_k \in R^{m\times m}$ the transfer matrix, and its (i, j) element can be given by

$$\mathbf{H}_k(i,j) = S_{ij}\left[\left(\beta_{ik}^2 - 1\right) + 2\mathbf{i}\zeta_i\beta_{ik}\right]^{-1}\left[\left(\beta_{jk}^2 - 1\right) - 2\mathbf{i}\zeta_j\beta_{jk}\right]^{-1} \tag{6}$$

and $\beta_{ik} = f_i/\mathrm{f}_k$; f_i is the natural frequency of the i-th mode; S_{ij} is the cross spectral density between the i-th and j-th modal excitation. The first and second term in Eq. (5) represent the contribution from the modal response and the prediction error, respectively.

Theoretically, the modal parameters can be determined by maximizing the posterior PDF. However, there are some computational difficulties. To develop an efficient algorithm, a fast algorithm was developed, and it allows the MPV and the associated posterior uncertainty to be obtained efficiently [16–18].

2.2. Bayesian method for multiple setups

In full-scale ambient tests, usually there are a large number of DOFs to be measured, but the number of available sensors is often limited. In this part, a fast Bayesian method for modal identification of well separated modes incorporating data from multiple setups will be presented. For the details of this part, please refer to [11, 19, 20].

Assume that in a selected frequency band, there is only one contributing mode and $\boldsymbol{\varphi} \in R^n$ denotes the global mode shape containing all the DOFs of interest. To relate the $\boldsymbol{\varphi}$ to the mode shape in a given setup covering a possibly different set of DOFs, the selection matrix in Setup i ($i = 1, \ldots, n_s$), $\mathbf{L}_i \in R^{n_i \times n}$ will be defined and n_i denotes the number of measured DOFs in Setup i. In Setup i, the (j,k) entry of $\mathbf{L}_i$ is set to 1 when DOF k is measured by the j-th channel, and it is equal to zero for the other situations. The theoretical mode shape confined to the measured DOFs in Setup i can be given by $\boldsymbol{\varphi}_i = \mathbf{L}_i\boldsymbol{\varphi} \in R^{n_i}$. The modal parameters to be identified are defined as

$$\boldsymbol{\theta} = [f_i, \zeta_i, S_i, S_{ei} : i = 1, \ldots, n_s; \boldsymbol{\varphi} \in R^n] \in R^{4n_s+n} \tag{7}$$

in which $f_i, \zeta_i, S_i, S_{ei} (i = 1, \ldots, n_s)$ denote the modal parameters in Setup i.

Let $\mathbf{Z}_k^{(i)} = [\mathrm{Re}\mathcal{F}_{ik}; \mathrm{Im}\mathcal{F}_{ik}] \in R^{2n_i} (i = 1, \ldots, n_s)$ denote the FFT data at frequency f_k in the i-th setup; $\mathcal{D}_i = \{\mathbf{Z}_k^{(i)}\}$ denote the FFT data in a selected frequency band in Setup i and $\mathcal{D} = \{\mathcal{D}_i : i = 1, \ldots, n_s\}$ denote the FFT data in all the setups. Assuming the data in different setups are independent, based on Bayes' Theorem, given the data in all setups, the posterior PDF of $\boldsymbol{\theta}$ is given by

$$p(\boldsymbol{\theta}|\mathrm{D}) \propto p(\{D_1, D_2, \ldots, D_{n_s}\}|\boldsymbol{\theta}) = p(D_1|\boldsymbol{\theta})p(D_2|\boldsymbol{\theta})\ldots p(D_{n_s}|\boldsymbol{\theta}) \tag{8}$$

Note that $p(\mathcal{D}_i|\boldsymbol{\theta})$ is independent with the modal parameters in other setups, and so $p(\mathcal{D}_i|\boldsymbol{\theta}) = p(\mathcal{D}_i|f_i, \zeta_i, S_i, S_{ei}, \boldsymbol{\varphi}_i)$. Eq. (8) can be given by

$$p(\boldsymbol{\theta}|\mathcal{D}) \propto \prod_{i=1}^{n_s} p(\mathcal{D}_i|f_i, \zeta_i, S_i, S_{ei}, \boldsymbol{\varphi}_i) \tag{9}$$

Similar to above section, $p(\mathcal{D}_i|f_i, \zeta_i, S_i, S_{ei}, \boldsymbol{\varphi}_i)$ is asymptotically a Gaussian distribution, with negative log-likelihood function (NLLF)

$$L_i(\boldsymbol{\theta}_i) = \frac{1}{2}\sum_k \left[\ln \det \mathbf{C}_{ik}(\boldsymbol{\theta}_i) + \mathbf{Z}_k^{(i)T}\mathbf{C}_{ik}(\boldsymbol{\theta}_i)^{-1}\mathbf{Z}_k^{(i)}\right] \tag{10}$$

in which $\boldsymbol{\theta}_i = \{f_i, \zeta_i, S_i, S_{ei}, \boldsymbol{\varphi}_i\}$ and

$$\mathbf{C}_{ik}(\boldsymbol{\theta}_i) = \frac{S_i D_{ik}}{2}\begin{bmatrix} \boldsymbol{\varphi}_i\boldsymbol{\varphi}_i^T & 0 \\ 0 & \boldsymbol{\varphi}_i\boldsymbol{\varphi}_i^T \end{bmatrix} + \frac{S_{ei}}{2}\mathbf{I}_{2n_i} \tag{11}$$

is the theoretical covariance matrix of the FFT data at the k-th frequency abscissa in Setup i, in which

$$D_{ik}(f_i, \zeta_i) = \left[(\beta_{ik}^2 - 1)^2 + (2\zeta_i\beta_{ik})^2\right]^{-1} \tag{12}$$

with $\beta_{ik} = f_i/f_k$. Consequently,

$$p(\boldsymbol{\theta}|\mathcal{D}) \propto \exp\left(-L(\boldsymbol{\theta})\right) \tag{13}$$

in which

$$L(\boldsymbol{\theta}) = \sum_{i=1}^{n_s} L_i(\boldsymbol{\theta}_i) \tag{14}$$

Similar to the case in single setup in Section 2.1., there are some computational difficulties. For the purpose of developing a fast computational procedure, it is found $\det\mathbf{C}_{ik}$ and $\mathbf{C}_{ik}^{-1}$ can be analysed and obtained in a more manageable form. As a result (proof omitted here), the overall NLLF can be reformulated as

$$\begin{aligned} L(\boldsymbol{\theta}) &= -(\ln 2)\sum_{i=1}^{n_s} n_i N_{fi} + \sum_{i=1}^{n_s}(n_i - 1)N_{fi}\ln S_{ei} \\ &+ \sum_{i=1}^{n_s}\sum_k \ln(S_i D_{ik}\|\mathbf{L}_i\boldsymbol{\varphi}\|^2 + S_{ei}) + \sum_{i=1}^{n_s} S_{ei}^{-1} d_i - \boldsymbol{\varphi}^T\mathbf{A}(\boldsymbol{\varphi})\boldsymbol{\varphi} \end{aligned} \tag{15}$$

where

$$\mathbf{A}(\boldsymbol{\varphi}) = \sum_{i=1}^{n_s} S_{ei}^{-1}\sum_k (\|\mathbf{L}_i\boldsymbol{\varphi}\|^2 + \frac{S_{ei}}{S_i D_{ik}})^{-1}\mathbf{L}_i^T\mathbf{D}_{ik}\mathbf{L}_i \in R^{n\times n} \tag{16}$$

$$\mathbf{D}_{ik} = \mathrm{Re}\mathcal{F}_{ik}\mathrm{Re}\mathcal{F}_{ik}{}^T + \mathrm{Im}\mathcal{F}_{ik}\mathrm{Im}\mathcal{F}_{ik}{}^T \in R^{n_i\times n_i} \tag{17}$$

$$d_i = \sum_k (\mathrm{Re}\mathcal{F}_{ik}{}^T\mathrm{Re}\mathcal{F}_{ik} + \mathrm{Im}\mathcal{F}_{ik}{}^T\mathrm{Im}\mathcal{F}_{ik}) \tag{18}$$

Based on Eq. (15), an iterative scheme for the full set of solution of modal parameters can be developed, which makes the identification of modal parameters very efficiently even for a large number of measured DOFs. The associated posterior uncertainties can be calculated by the analytical formulas without resorting to finite difference. Details can be found in Refs. [19, 20].

3. Two super tall buildings in Hong Kong

Buildings A and B are two super tall buildings in Hong Kong. Building A is 310 m tall, 50 m by 50 m in plan, which is a tabular concrete building with a central core wall system. Building B is 320 m tall, 50 m × 50 m in plan, whose lateral structural resistance is provided by two outrigger trusses with core walls near the centre and mega columns at the corner. **Figure 1** shows the mode

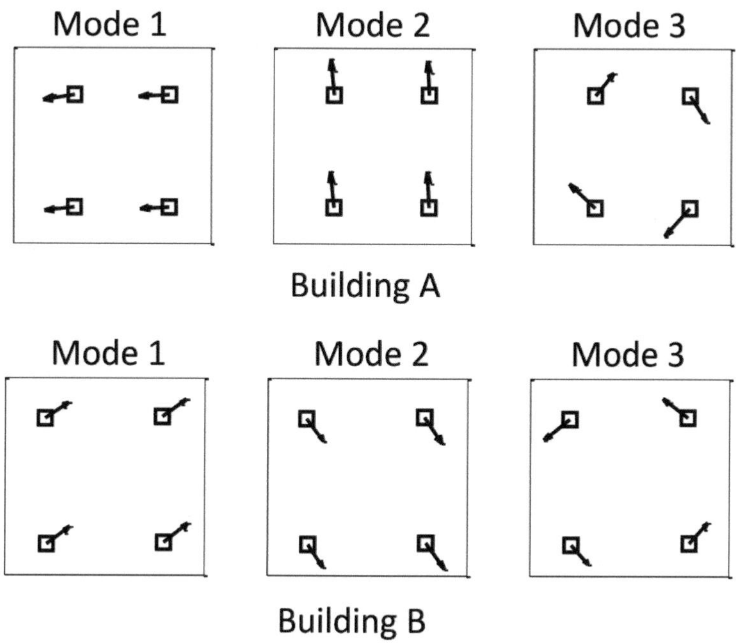

Figure 1. Mode shapes of Buildings A and B identified under normal wind.

shapes of the first three modes of Buildings A and B. They were identified based on ambient data by placing four tri-axial accelerometers at four corners on the roof. The two numbers near to the mode name are the natural frequency and damping ratio, respectively. For Building B, due to connection to a neighbouring building, the principal modal directions do not align with the building sides.

Figure 2 shows the sensor layout inside the rooms on the roof of the buildings, and the sensor locations were kept the same for all the typhoon events. The vibration data investigated were measured during two typhoon events (namely, Typhoon Goni and Koppu) and two monsoon events (MS1 and MS2). Typhoon Goni attacked Hong Kong during 4–6 August 2009. The wind speed changed between 22 and 60 km/hr. One month later, Typhoon Koppu attacked Hong Kong. Koppu travelled faster than Goni. The wind speed changed between 25 and 120 km/hr. Typhoon Koppu provides much more information due to large structural vibrations in the current study. The MS1 visited during 4–7 January 2010. The wind speed changed between 15 and 50 km/hr. The MS2 visited in December 2010. The wind speed ranged between 12 and 45 km/hr. The wind speeds in the two monsoon events are significantly lower than those in the typhoon events. The typhoon events can provide information for large amplitude vibration study, while the monsoon events can provide information in the low to moderate wind speed regime.

The recorded acceleration data are used for identifying the modal parameters of the structures by the Bayesian method. For the purpose that the loading and the response can be modelled as stationary stochastic process, the whole time history is divided into non-overlapping time windows of 30 minutes. **Figure 3** shows the singular value (SV) spectra of the first half hour data of Building A during Typhoon Koppu. Six modes below 1 Hz can be observed while the

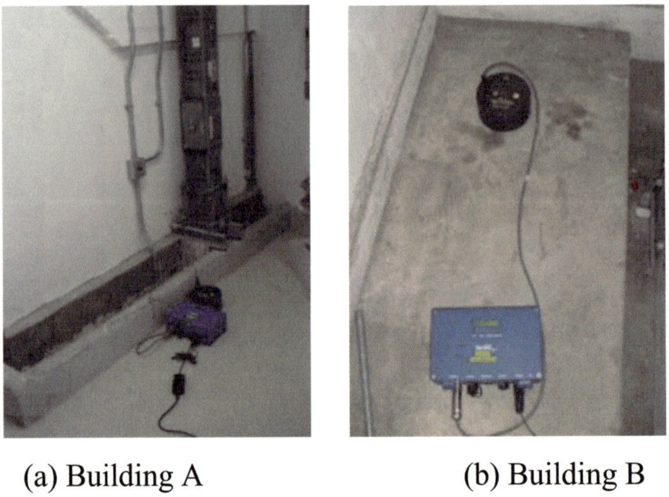

Figure 2. Equipment used during strong wind event.

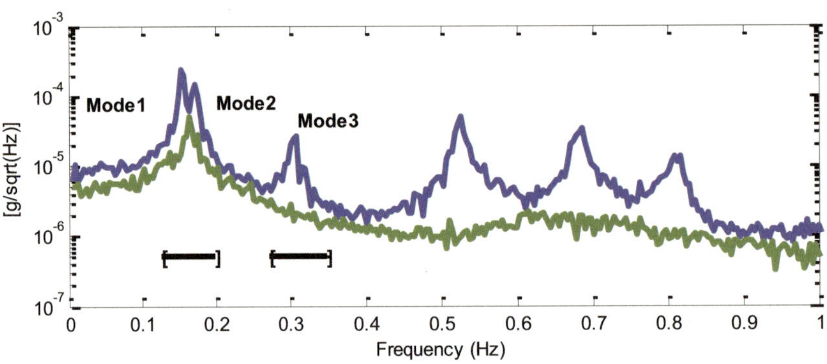

Figure 3. Root singular value spectra, 30-minute acceleration data, Building A.

first two modes are closely spaced. Similarly, in Building B, whose SV spectra is omitted here, there are also six modes below 1 Hz and the first two modes are closely spaced modes. The closely spaced modes may be due to their square-shaped floor plan of Buildings A and B. The first three modes are investigated including the first two translational modes and the first torsional mode.

Table 1 shows the modal parameters for Building A. The MPV of modal parameters and the corresponding posterior coefficient of variations (c.o.v.) are provided in the table. The posterior c.o.v.s of the natural frequencies are all less than 1%, which are much smaller than those of the damping ratio or the PSD of modal force. The posterior uncertainty of the damping ratio governs the data length requirement and with 30 minutes of data its c. o.v. is about 30%, showing a moderate level of accuracy. Similar investigation to Building B is also performed.

http://dx.doi.org/10.5772/intechopen.68397

Next, the correlation between the identified natural frequencies and damping ratios with the vibration amplitude calculated based on Ref. [6] is investigated. **Figures 4** and **5**, respectively, show the MPV of natural frequencies and damping ratios versus the modal vibration root-mean-square (RMS) for Buildings A and B. The results for all time windows in four different events are plotted in the figure. A circle, a cross and a triangle denote the results for Typhoon Goni, Typhoon Koppu and MS2, respectively, in Building A, while for Building B, a cross and a

Mode	f **(Hz)**		ζ **(%)**		S $[(\mu g)^2/Hz]$	
	MPV	**(c.o.v.)**	**MPV**	**(c.o.v.)**	**MPV**	**(c.o.v.)**
TX1	0.153	(0.16%)	0.5	(36%)	225.4	(9.0%)
TX2	0.170	(0.31%)	1.7	(20%)	349.1	(10%)
R1	0.302	(0.17%)	0.9	(20%)	4.5	(10%)

Table 1. Summary of identified modal properties, building A, first 30 minutes in Koppu.

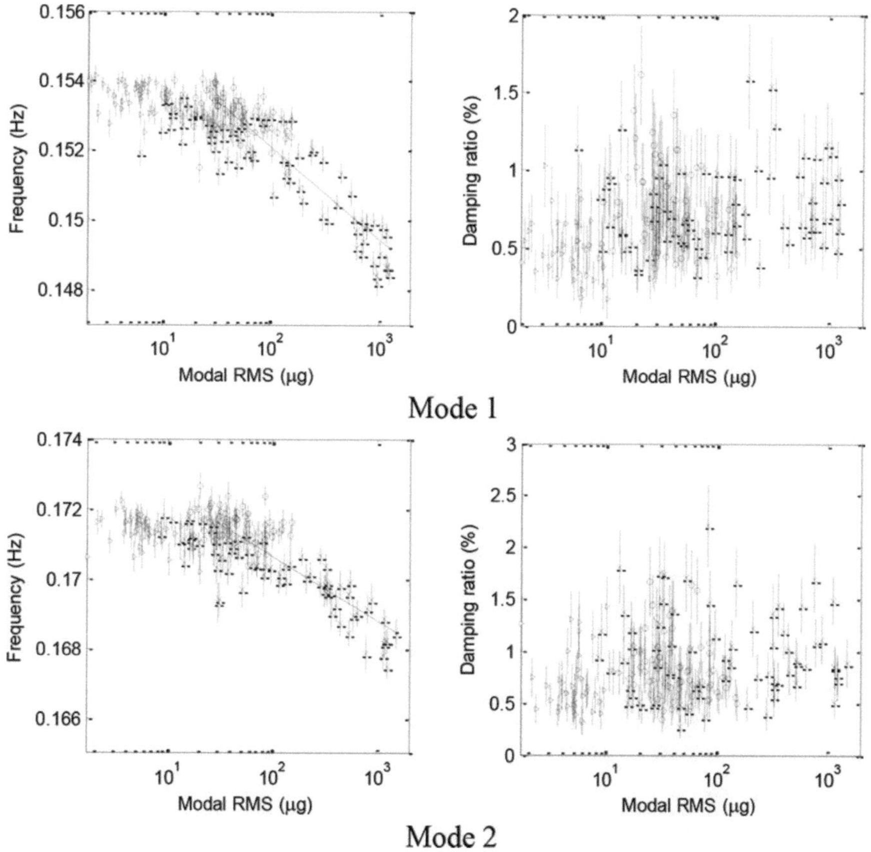

Figure 4. Identified modal frequency and damping ratio versus modal RMS of Building A (o: Goni; ×: Koppu; ⊳: MS2).

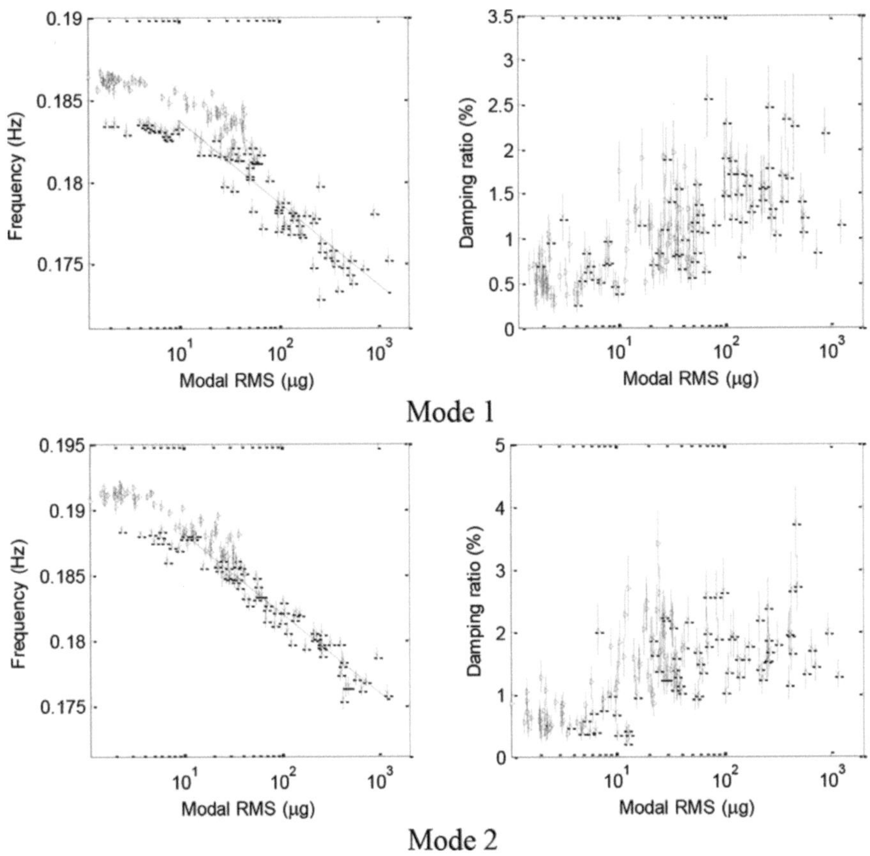

Figure 5. Identified modal frequency and damping ratio versus modal RMS of Building B (×: Koppu, ⊳: MS1).

triangle, respectively, denote the result for Typhoon Koppu and MS1. In the figures, each point is centred at the MPV with an error bar covering +/– one posterior standard derivation of the modal parameter. It is obvious from that the scatter in the MPV is significantly larger than those implied by the error bars, indicating some systematic dependence on the vibration amplitude.

There is an inverse trend between the natural frequency and the RMS value of the modal response, regardless of mode and building. Due to a season effect, the identified natural frequencies in Koppu (×) (in summer) are not the same with those in MS1 (⊳) (in winter) at low modal RMS. The natural frequency tends to decrease with the increase of modal RMS. One possible reason is the loosening of friction joints at sufficiently high vibration levels, which results in the reduction of stiffness. The right column of **Figures 4** and **5** shows a positive correlation between the damping ratio and RMS. Compared to that in the frequencies, the scatter is much bigger. The scatter in these figures can be due to limited identification precision, modelling error in the damping mechanism, modelling error in stationarity, unknown amplitude-dependence mechanism, etc. For the more detailed information of the study about the two super tall buildings, please refer to [6].

4. Canton Tower

The Canton Tower as shown in **Figure 6** situated in Guangzhou, China, is a super tall structure with a height of 610 m. It is composed of two tube-like structures, i.e., a reinforced concrete inner structure and a steel outer structure with concrete-filled-tube (CFT) columns.

The SHM system deployed on the Canton Tower is composed by more than 700 sensors, including anemometers, accelerometers, fibre optic strain and temperature sensors, global position system, and so on [24, 25]. Among them, 20 uni-axial accelerometers are installed on eight different cross-sections of the inner structure. On cross-sections 4 and 8, four accelerometers are instrumented at two locations for bi-axial measurement; while on each of the other cross-sections, two sensors are installed at two locations for uni-axial measurement. Sensors 01, 03, 05, 07, 08, 11, 13, 15, 17, 18 are deployed to collect the structural response in the short-axis direction, while sensors 02, 04, 06, 09, 10, 12, 14, 16, 19, 20 measure the structural response in the long-axis direction. The frequency range of accelerometers is DC to 50 Hz and amplitude range is ± 2 g. The 24-hour acceleration data were collected from 18:00, 20th January 2010 to 18:00, 21st January 2010 under normal wind condition. The sampling frequencies of acceleration are set to be 50 Hz.

As the last example, the 24-hour data were separated into 48 time windows with 30 minutes for each window. **Figure 7** shows the root PSD spectra of the structural responses of a typical

Figure 6. Overview of Canton Tower.

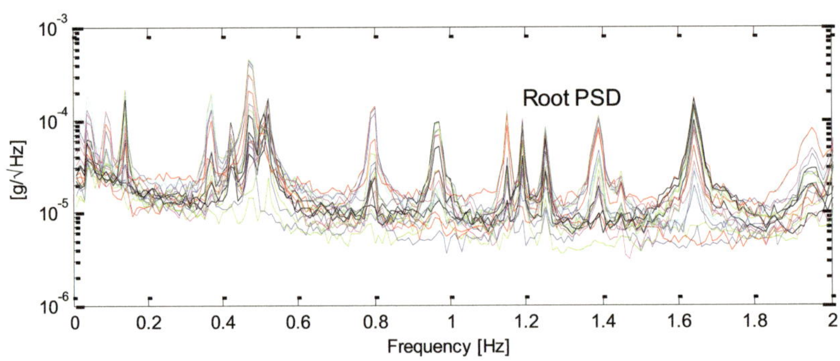

Figure 7. Power spectral density for a typical 30-minute window.

time window. Fifteen peaks can be observed below 2.0 Hz. The first mode is less than 0.1 Hz, and it is the foundational mode of the Canton Tower. In the operational modal analysis by the Fast Bayesian FFT method, all the 15 modes are identified. **Table 2** shows the identified modal parameters of the 15 modes. From the second to the ninth column, every two columns are considered as one group with first denoting the MPV and second denoting the associated posterior c.o.v. The c.o.v. of modal frequencies are quite small (less than 0.5%), implying that the MPVs are quite accurate. The damping ratios for this structure are small and only those of the first and fourth modes are higher than 1%. This is consistent with the results obtained in Ref. [24]. The posterior uncertainty of damping ratios is relatively high in comparison with that

Mode	Characteristics	f (Hz)	COV (%)	ζ (%)	COV (%)	S (μg²/Hz)	COV (%)	Se (μg²/Hz)	COV (%)
1	Bending	0.094	0.37	1.20	32.5	1.64	17.5	5.56	3.21
2	Bending	0.138	0.18	0.48	39.9	0.92	17.3	3.49	3.50
3	Bending	0.366	0.08	0.26	32.5	0.30	10.9	1.49	2.28
4	Bending	0.424	0.07	0.21	35.7	0.05	17.1	2.25	3.28
5	Bending	0.475	0.05	0.12	41.6	0.88	11.8	1.77	2.58
6	Torsion	0.506	0.04	0.10	46.4	0.05	21.5	8.83	4.00
7	Bending	0.522	0.07	0.27	29.1	0.19	15.6	2.52	3.01
8	Bending	0.796	0.05	0.23	23.2	0.08	7.8	0.81	1.61
9	Bending	0.966	0.06	0.36	17.5	0.06	7.6	0.65	1.50
10	Combined	1.151	0.03	0.13	26.8	0.01	11.7	0.63	2.35
11	Bending	1.191	0.03	0.11	29.1	0.01	12.9	0.84	2.53
12	Torsion	1.250	0.03	0.11	27.6	0.01	9.8	0.66	1.94
13	Bending	1.388	0.05	0.33	16.0	0.05	8.4	0.50	1.63
14	Bending	1.643	0.04	0.22	16.7	0.07	6.4	0.51	1.33
15	Combined	1.946	0.08	0.74	11.6	0.09	10.3	0.62	1.53

f: modal frequency; ζ: damping ratio; S: PSD of modal force; Se: PSD of prediction error.

Table 2. Identified modal parameters and the associated posterior uncertainty.

of modal frequencies, with an order of magnitude of a few tens' percent. The PSD of modal force and PSD of prediction error are all related to the excitation environment. The c.o.v. of the former is apparently larger than the c.o.v. of the latter.

Figures 8 and **9** show the identified mode shapes of the fifteen modes projected in short- and long-axis directions, respectively. As aforementioned, on cross-sections 4 and 8, four accelerometers were deployed for bi-axial measurement at two plane locations. With this information, the torsional behaviour of these modes can be investigated. From the mode shapes identified (omitted here), although the mode shapes of some modes in **Figures 8** and **9** are similar when projected in short- or long-axis direction, they are different in top view, i.e. Modes 6, 10, 12, 15 exhibit a significant torsional behaviour.

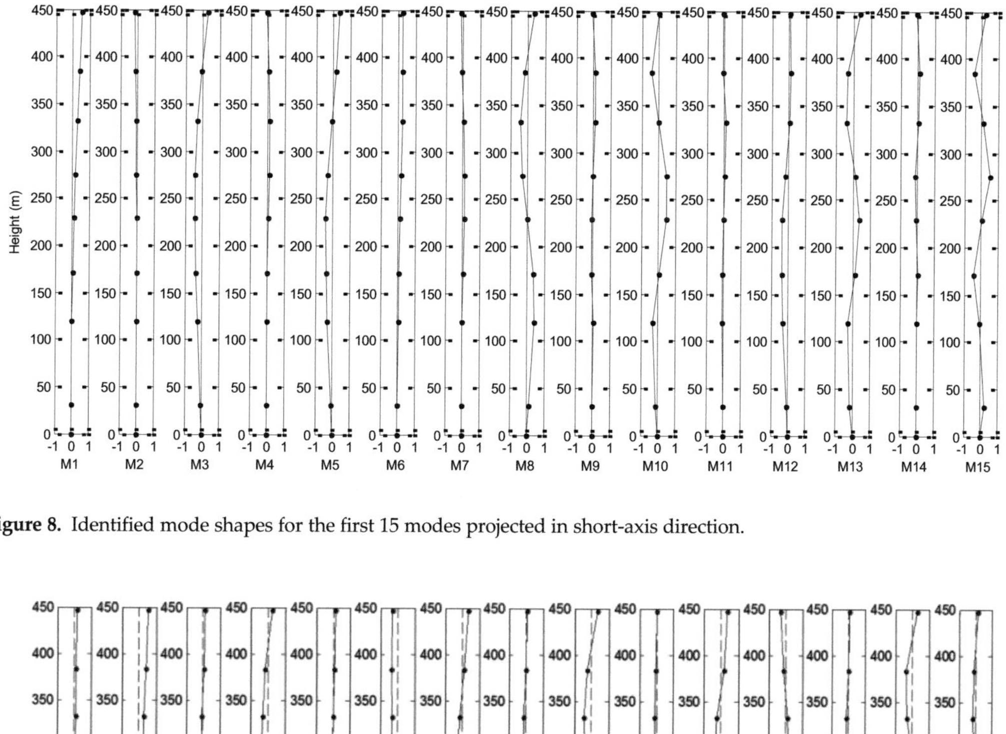

Figure 8. Identified mode shapes for the first 15 modes projected in short-axis direction.

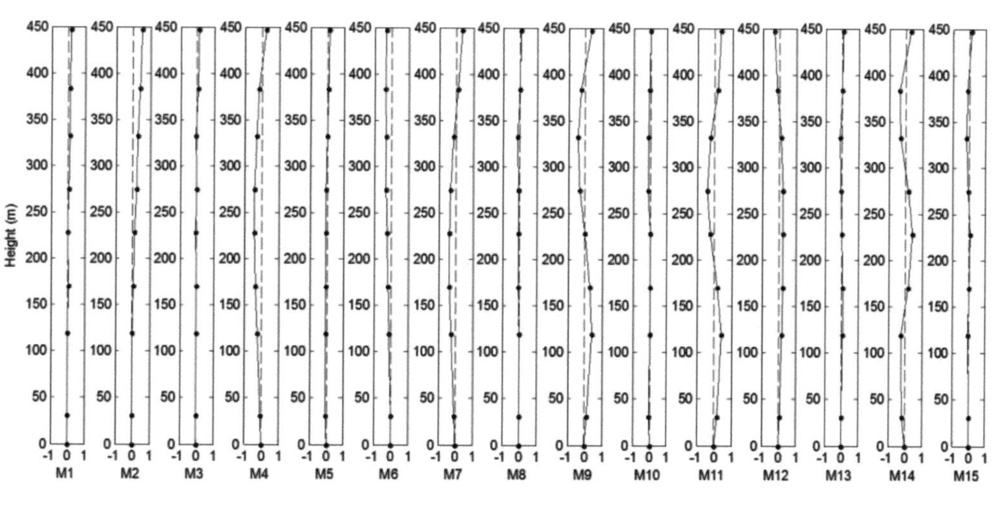

Figure 9. Identified mode shapes for the first 15 modes projected in long-axis direction.

To verify the results identified by the Fast Bayesian FFT method, the enhanced FDD method is used to identify the modal parameters using the same data. **Table 3** shows the identified modal frequencies and damping ratios. The identified modal frequencies for different modes are consistent with their counterparts identified by the Fast Bayesian FFT method. However, there is a larger discrepancy for the damping ratios. A noticeable difference can be observed between the two groups of results, implying that high uncertainty exists in the identified damping ratios, as shown by the large posterior c.o.v. of damping ratios obtained by the Fast Bayesian FFT method. For the more detailed information of the study about Canton Tower, please refer to [4].

Mode	Modal frequency (Hz)	Damping ratio (%)
1	0.094	2.48
2	0.138	1.33
3	0.366	0.46
4	0.424	0.32
5	0.475	0.29
6	0.506	0.28
7	0.522	0.43
8	0.796	0.46
9	0.965	0.64
10	1.151	0.16
11	1.191	0.16
12	1.250	0.16
13	1.390	0.34
14	1.642	0.27
15	1.948	0.86

Table 3. Modal parameters identified by enhanced FDD method.

5. Shanghai Tower

The Shanghai Tower (**Figure 10**) is a 124-story 632 m high super tall structure, situated in Lujiazui, Shanghai, China. The structure has eight electromechanical floor zones with six two-storey outrigger trusses together with eight boxy space circular trusses set along these different zones. It has a mega frame-tube-outrigger lateral resistant system, where the mega frame is composed by the boxy space circular truss and the giant column. To monitor the structural condition of the super tall building, an SHM system was instrumented on this structure. Many kinds of sensors were installed including accelerometers, temperature sensors, GPS, etc. A series of field vibration tests were conducted to investigate its modal parameters.

Figure 10. Overview of the building.

5.1. Field vibration tests in different stages

From April 2012 to December 2014, to study the modal parameters during construction, 15 ambient vibration tests were carried out over a period of two and a half years. **Table 4** shows the time to carry out the field test corresponding to the number of floors constructed. Two different locations were measured in each test. They were at the top of a core tube and the top of composite slabs after completion of concrete pouring. A series of finite element models (FEMs) (built by ETABS) of the first eight construction stages were also developed to perform comparison. The measured structures and the FEM model are shown in **Figure 11**.

When the main structure was finished, one field test was conducted to measure different corners of the tube to investigate the dynamic characteristics of a typical floor. For the convenience of sensor alignment and cables arrangement, 101th floor was selected since only the shear walls were constructed on this floor. It was planned to measure nine locations bi-axially,

Setup	1	2	3	4	5	6	7	8	9	10	11	12	13	14	15
Year/Month	12/04	12/07	12/08	12/10	13/01	13/03	13/05	13/07	13/08	13/12	14/02	14/03	14/07	14/10	14/12
Floors	55	68	71	81	94	102	111	120	125	125	125	125	125	125	125

Table 4. Measurement information.

Figure 11. Overview of buildings in different stages and finite element models, field test on a typical floor.

which are shown in **Figure 12**. It includes one location at the centre and the locations in the eight corners. With the help of core walls, sensor alignment was finished in half an hour.

In this test, only four uniaxial sensors were available, and so to finish the whole measurement, multiple setups were designed. Two reference channels were put in Location 1 to provide common information for mode shapes assembling and they were kept unchanged during the whole measurement. Based on the number of sensors and locations, eight different setups were

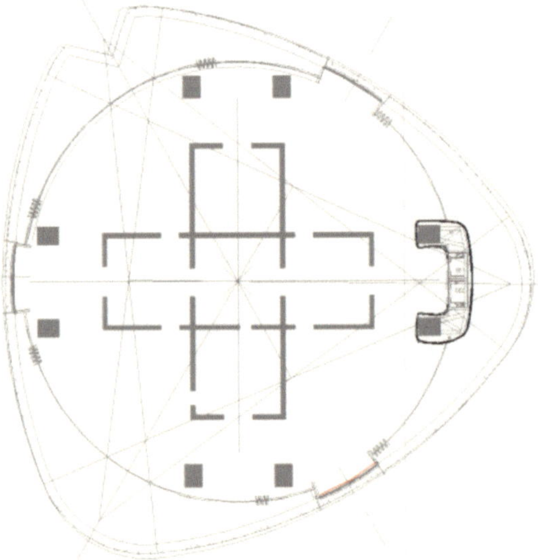

Figure 12. Setup plan.

arranged. The plans for these setups can be found in Ref. [26]. Forty minutes are required at least for each setup. Thirty minutes were used for data collection and the remaining 10 minutes were used for roving the sensor. It covered 10 am to 7 pm during a working day to finish the whole measurement with the sampling frequency of 2048 Hz. For the convenience of analysis, the measured data were decimated to a sampling frequency of 64 Hz.

5.2. Data analysis

Using the data collected in different construction stages, the variations of modal parameters were investigated. In each test, 20 minutes data in the Location 2 were analysed. **Figure 13** provides the identified results of natural frequencies and damping ratios for Modes 1 and 2. The MPV is denoted by a dot, while ±2 posterior standard deviations were expressed by an error bar. The MPV of natural frequencies decreases with the number of floors and the increasing speed tends to be stable after the main structure was finished. The posterior uncertainty of the natural frequency is small, and so the decrease of the natural frequency with the number of floors constructed is due to the structural height instead of identification error since the error bars among neighbouring setups have no overlap. For the damping ratios, the MPVs are all around or less than 1% and no obvious trend can be observed with the structural height. The posterior uncertainty of damping ratios is relatively large, which can be seen in **Figure 13**.

The identified results of the first two modes and the corresponding ones obtained from the FEM are shown in **Figure 14**. From the figure, it is seen that the two group results were consistent with each other, which implies that the influence of the increase of structural height

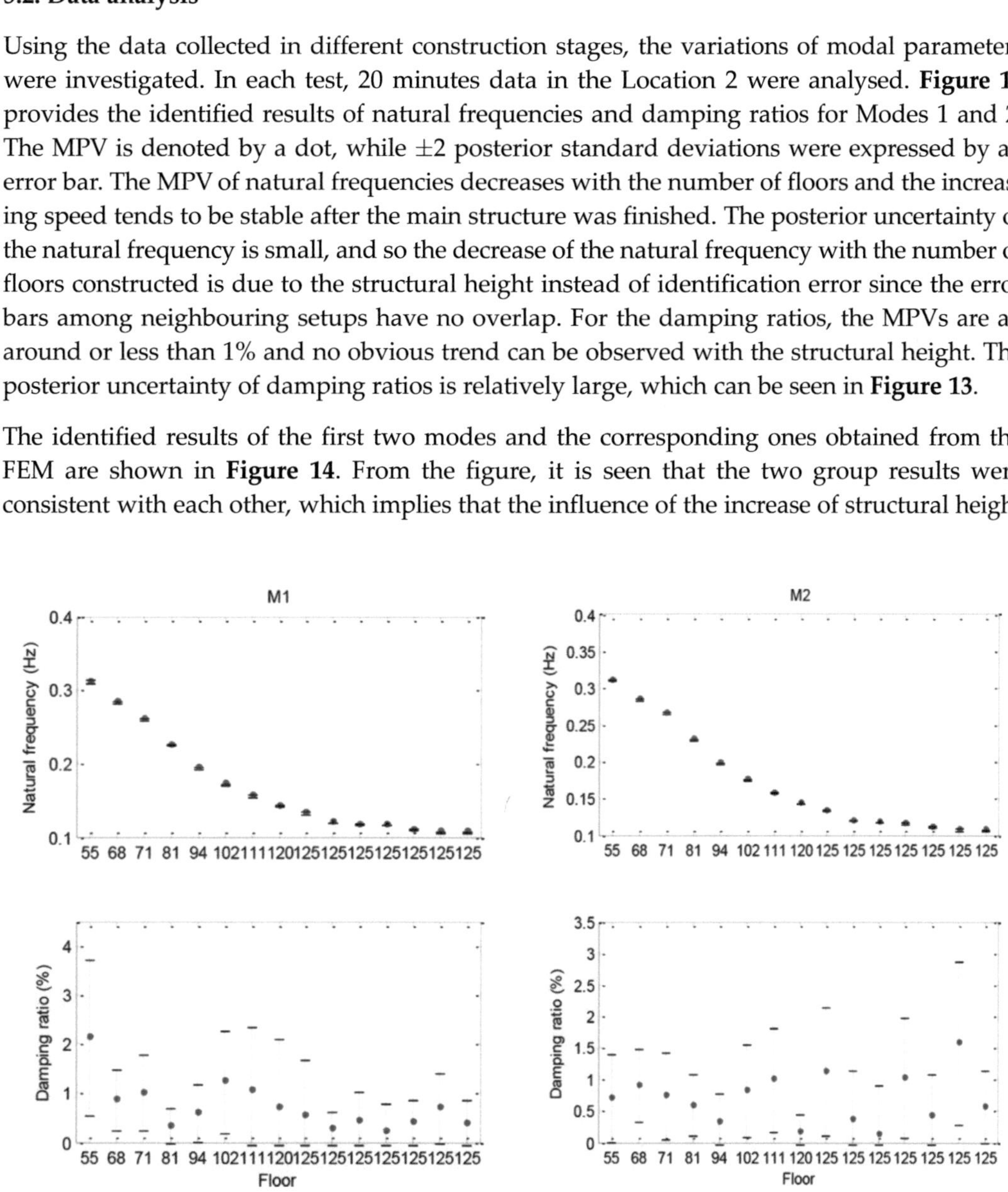

Figure 13. Modal parameters in different construction stages.

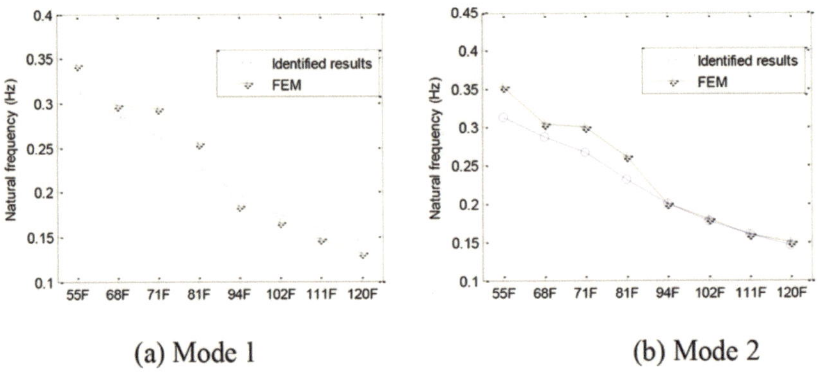

Figure 14. Comparison between the identified results and FEM results. (a) Mode 1 and (b) Mode 2.

on the reduction of natural frequency is reasonable. No obvious problem from the point of view of this modal parameter can be observed during construction.

Next, the data of the field test in a typical floor will be investigated. **Figure 15** plots the PSD spectra of the data in Setup 1. Eight obvious peaks can be observed from 0 to 1 Hz. The numbers near to each peak indicate the potential modes. There are two closely spaced modes (Modes 1 and 2) whose mode shapes can be predicted by the FEM model to be two transitional modes in x- and y-directions of the building, respectively. Therefore, to use the proposed multiple setups algorithm, the data in x- and y-directions were analysed separately to obtain the mode shapes. For Mode 3 to Mode 8, they can be taken as well separated, and so they are identified directly using the collected data.

The mode shapes of the first three identified modes were shown in **Figure 16**. Modes 1 and 2 are, respectively, two translational modes along the x- and y-directions. The third mode is a torsional mode and its torsion centre is at the centre of the tube. The mode shapes of Modes 4–6 are similar to Modes 1–3. Modes 7 and 8 (omitted here) are also, respectively, the two translational modes in the x- and y-directions. From the mode shapes, it is observed that although this super building was designed in a novelty manner, the mode shapes of the first eight modes are still all regular,

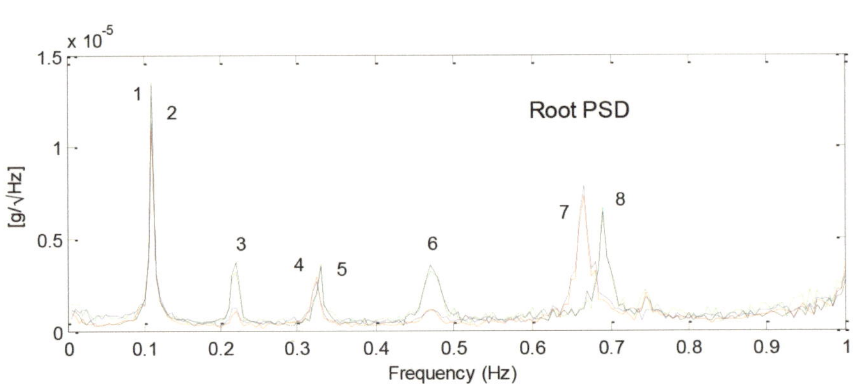

Figure 15. PSD spectrum of the data in Setup 1: (a) PSD spectra and (b) SVD spectra.

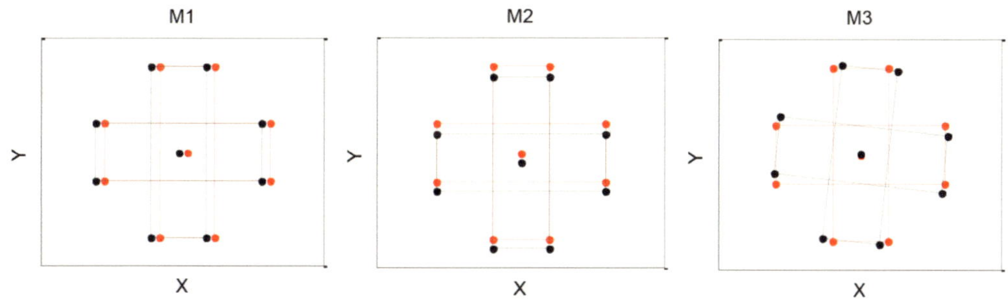

Figure 16. Identified mode shapes: Mode 1, Mode 2, and Mode 3.

i.e. two translational modes and then one torsional mode. **Table 5** gives the averaged posterior c.o.v. and the sample c.o.v. (=sample standard derivation/sample mean) of identified modal parameters for the first eight modes among different setups. The posterior c.o.v. tends to be larger than the sample c.o.v., while their orders of magnitude are still similar. These two quantities are well consistent with each other, although they can, respectively, reflect Bayesian and frequentist perspectives. For the more detailed information of the study about Shanghai Tower, please refer to [26].

Mode		1	2	3	4	5	6	7	8
f(%)	Pc.o.v.	0.37	0.34	0.23	0.18	0.35	0.14	0.18	0.21
	Sc.o.v.	0.48	0.37	0.27	0.17	0.40	0.21	0.12	0.67
z(%)	Pc.o.v.	57	46	38	48	48	25	26	32
	Sc.o.v.	88	48	53	41	62	31	27	33

Table 5. Posterior c.o.v. and sample c.o.v..

6. Conclusion

This chapter presents the work on the operational modal analysis of four super tall buildings including two super tall buildings situated in Hong Kong, Canton Tower and Shanghai Tower. A fast Bayesian method is used to perform the OMA. It is found that the Bayesian method can be well applied into these four field structures. The natural frequencies of the first fundamental mode of these four buildings are around 0.1 Hz, while the damping ratios are all around 1%. In addition to the most probable values of modal parameters, the associated posterior uncertainties are also investigated. The posterior c.o.v. of natural frequencies are usually small, indicating the identification of this quantity is accurate, while that of damping ratios are obviously larger than the natural frequencies. This is consistent with the common finding. The investigation in this chapter provides a reference for future OMA of super tall buildings, and the Fast Bayesian FFT method is a robust method having the potential to be used in other field structures.

Acknowledgements

The work in this paper was partly supported by grants from National Natural Science Foundation of China through Grant 51508413 and 51508407, Shanghai Pujiang Program (Grant No.: 15PJ1408600) and Fundamental Research Funds for the Central Universities (Grant No.:20161143).

Author details

Feng-Liang Zhang and Yan-Chun Ni*

*Address all correspondence to: yanchunni@gmail.com

College of Civil Engineering, Tongji University, Shanghai, China

References

[1] Chang PC, Flatau A, Liu SC. Health monitoring of civil infrastructure. Structural Health Monitoring. 2003;**2**(3):257–267

[2] Van der Auweraer H, Peeters B. International research projects on structural health monitoring: An overview. Structural Health Monitoring. 2003;**2**(4):341–358

[3] Ni YQ, Xia Y, Liao WX, Ko JM. Technology innovation in developing the structural health monitoring system for Guangzhou New TV Tower. Structural Control and Health Monitoring. 2009;**16**:73–98

[4] Zhang FL, Ni YQ, Ni YC, Wang YW. Operational modal analysis of Canton Tower by a fast frequency domain Bayesian method. Smart Structures and Systems. 2016;**17**(2):209-230

[5] Kijewski-Correa T, Kwon DK, Kareem A, Bentz A, Guo Y, Bobby A, Abdelrazaq A. Smartsync: An integrated real-time structural health monitoring and structural identification system for tall buildings. Journal of Structural Engineering, ASCE. 2013;**139** (10):1675–1687

[6] Au SK, Zhang FL, To P. Field observations on modal properties of two tall buildings under strong wind. Journal of Wind Engineering and Industrial Aerodynamics. 2012;**101**:12–23

[7] Li QS, Yi J. Monitoring of dynamic behaviour of super-tall buildings during typhoons. Structure and Infrastructure Engineering: Maintenance, Management, Life-Cycle Design and Performance. 2016;**12**(3):289–311

[8] Lam HF, Hu J, Yang JH. Bayesian operational modal analysis and Markov chain Monte Carlo-based model updating of a factory building. Engineering Structures. 2017;**132**: 314–336

[9] Au SK, Zhang FL. Fundamental two-stage formulation for Bayesian system identification, Part I: General theory. Mechanical Systems and Signal Processing. 2016;**66**:31–42

[10] Zhang FL, Au SK. Fundamental two-stage formulation for Bayesian system identification, Part II: Application to ambient vibration data. Mechanical Systems and Signal Processing. 2016;**66**:43–61

[11] Au SK, Zhang FL, Ni YC. Bayesian operational modal analysis: Theory, computation, practice. Computers and Structures. 2013;**126**:3–14

[12] Peeters B, De RG. Stochastic system identification for operational modal analysis: A review. Journal of Dynamic Systems Measurement and Control-Transactions of the ASME. 2001;**123**(4):659–667

[13] Brincker R, Zhang L, Anderson P. Modal identification of output-only systems using frequency domain decomposition. Smart Materials and Structures. 2001;**10**:441–455

[14] Yuen KV. Bayesian Methods for Structural Dynamics and Civil Engineering. New York: Wiley; 2010

[15] Yan WJ, Katafygiotis LS. A two-stage fast Bayesian spectral density approach for ambient modal analysis. Part I: most probable values and posterior uncertainty. Mechanical Systems and Signal Processing. 2015;**54**:139–155

[16] Au SK. Fast Bayesian FFT method for ambient modal identification with separated modes. Journal of Engineering Mechanics, ASCE. 2011;**137**:214–226

[17] Au SK. Fast Bayesian ambient modal identification in the frequency domain, Part I: Posterior most probable value. Mechanical Systems and Signal Processing. 2012;**26**:60–75

[18] Au SK. Fast Bayesian ambient modal identification in the frequency domain, Part II: posterior uncertainty. Mechanical Systems and Signal Processing. 2012;**26**:76–90

[19] Au SK, Zhang FL. Fast Bayesian ambient modal identification incorporating multiple setups. Journal of Engineering Mechanics, ASCE. 2012;**138**(7):800–815

[20] Zhang FL, Au SK, Lam HF. Assessing uncertainty in operational modal analysis incorporating multiple setups using a Bayesian approach. Structural Control and Health Monitoring. 2015;**22**(3):395–416

[21] Au SK, Ni YC. Fast Bayesian modal identification of structures using known single-input forced vibration data. Structural Control and Health Monitoring. 2014;**21**(3):381–402

[22] Zhang FL, Ni YC, Au SK, Lam HF. Fast Bayesian approach for modal identification using free vibration data, Part I - Most probable value. Mechanical Systems and Signal Processing. 2016;**70-71**:209–220

[23] Ni YC, Zhang FL, Lam HF, Au SK. Fast Bayesian approach for modal identification using free vibration data, Part II - Posterior uncertainty and application. Mechanical Systems and Signal Processing. 2016;**70-71**:221–244

[24] Chen WH, Lu ZR, Lin W, Chen SH, Ni YQ, Xia Y, Liao WY. Theoretical and experimental modal analysis of the Guangzhou New TV Tower. Engineering Structures. 2011;**33**:3628–3646

[25] Ni YQ, Xia Y, Lin W, Chen WH, Ko JM. SHM benchmark for high-rise structures: A reduced-order finite element model and field measurement data. Smart Structures and Systems. 2012;**10**(4):411–426

[26] Zhang FL, Xiong HB, Shi WX, et al. Structural health monitoring of Shanghai Tower during different stages using a Bayesian approach. Structural Control and Health Monitoring. 2016;**23**(11):1366–1384

Chapter 5

Mooring Integrity Management: Novel Approaches Towards *In Situ* Monitoring

Ángela Angulo, Graham Edwards, Slim Soua and Tat-Hean Gan

Additional information is available at the end of the chapter

http://dx.doi.org/10.5772/intechopen.68386

Abstract

The recent dramatic fluctuations in oil and gas prices are forcing operators to look at radically new ways of maintaining the integrity of their structures. Moreover, the life of old structures has to be extended. This includes the replacement of expensive periodic in-service inspections with cost-efficient structural health monitoring (SHM) with permanently installed sensors. Mooring chains for floating offshore installations, typically designed for a 25-year service life, are loaded in fatigue in a seawater environment. There is no industry consensus on failure mechanisms or even defect initiation that mooring chains may incur. Moorings arc safety-critical areas, which by their nature are hazardous to inspect. Close visual inspection in the turret is usually too hazardous for divers, yet is not possible with remotely operated vehicles (ROVs), because of limited access. Conventional non-destructive techniques (NDTs) are used to carry out inspections of mooring chains in the turret of floating production storage and offloading (FPSO) units. Although successful at detecting and assessing the fatigue cracks, the hazardous nature of the operation calls for remote techniques that can be applied continuously to identify damage initiation and progress. Appropriate replacement plans must enhance current strategies by implementing real-time data retrofit.

Keywords: mooring chain, structural integrity, structural health monitoring, acoustic emission, guided wave, crack growth

1. Introduction

As offshore exploration and production goes further afield and into deeper waters, more offshore operations, e.g. oil and gas operations [1], are conducted from floating platforms moored to the seabed by chains. Mooring lines are safety-critical systems on offshore floating and semi-submersible platforms. The lines are often subject to immense environmental and structural forces from currents, waves and hurricanes. Other forces arise from impact with the seabed, abrasion, increased drag by accumulation of marine organisms and salt water corrosion. Failure of one or more of these mooring lines can result in disastrous and economic consequences for safety, the environment and production.

Periodic inspection of chain systems is mandatory [2] for safety and early detection of faults and is usually performed either through:

- an in-air (outside the water) process that necessitates the removal of the chain for inspection at the surface. Although common practice with movable jack-up drilling rigs, for example, it is not possible with fixed production systems except by taking the system out of production;
- in-water inspection, which is carried out with the chain system *in situ*.

An advantage of in-water inspection being carried out *in situ* is that it is easy to identify which parts of the chain have been in the wear zone, i.e. in the thrash zone and at the fairlead. This is normally more difficult to determine for long lengths of chain inspected on the quayside [3]. However, the most important advantage of in-water inspection is that it can be carried out during the daily production by the facility with minimal stoppage time, hence excluding the need to decommission and the cost to business through lost production.

Currently, the volumetric non-destructive testing (NDT) used in the offshore industrial sector cannot be deployed underwater with the same efficiency without radical adaptation of the technology. In general, current in-water testing techniques have intrinsic issues with probability of detection that are amplified when applied underwater; hence, their use on mooring chains requires highly specialised procedures. This is limited:

- NDT diver-inspections are in general not a favoured option due to health and safety issues, inconsistency of results and an inherent depth limitation and risk, e.g. when checking the thrash zone;
- standalone robotic systems are too large and cumbersome for practical offshore operations. They are not able to inspect the chains in the thrash zone or near the chain fairleads;
- tethered remotely operated vehicles (ROVs) that use both mechanical and optical calliper systems have met with limited success primarily due to their method of deployment on the chain, i.e. they need in-water diver supervision as they have the potential to be knocked out of true positioning and must be recalibrated between successive measurements.

These limitations compromise the capability for early detection of faults, resulting in periodic failures. For example, between 1980 and 2001, the HSE reported [4] that a drilling semi-submersible in the North Sea can expect to experience a mooring failure (anchor dragging, breaking of mooring lines, loss of anchor, winch failures) once every 4.7 operating years, a production semi-submersible once every 9 years and a floating production storage and offloading (FPSO) installation once every 8.8 years, due to failure to detect defects in the chain.

http://dx.doi.org/10.5772/intechopen.68386

Consequently, reliance must be placed on in-air inspection, resulting in disruption to the daily operations and substantial economic loss for the operating companies, as the production structures require partial/full de-commissioning during the inspection period (**Figure 1**).

On another front, greater demand for energy in Europe [5], depletion of onshore resources and insecurity of supply from geopolitically unstable traditional areas [6], has led to a push for offshore oil and gas exploration in deep water, with substantial interest in marginal production fields. This necessitates floating production systems with massive mooring chain systems to overcome the substantial challenge for economic extraction. The reliability of inspection is dictated by three important factors [7] as follows:

i. the responsible operators must specify their requirements very clearly in terms of the regions to be inspected and the types of flaws or damage mechanism present (any decommissioning would be prohibitively expensive);

ii. the NDT methods, equipment and personnel must be proven to meet the purpose of the inspection through a suitable qualification process;

iii. the selected NDT process must be implemented thoroughly and mitigate the risk of failure, leading to reduction of scrap, repair, rework and workforce, and increased productivity and safety [8].

Figure 1. From left to right: Example of wear and corrosion on a chain link from the sea-bed touch down zone, in-water inspection showing a studded chain which has lost its stud *in situ*, illustration of marine growth on long-term deployed chain, affecting optical in-water inspection, friction-induced bending.

1.1. Damage mechanisms

The life cycle of a mooring system is in excess of 20 years, and it would normally be designed to withstand '100-year period storm' conditions. A typical floating structure has 14 moorings which can amount to nearly 10 km of chain or hybrid chain and polyester rope (central section). Mooring chains are subjected to cyclical loads and therefore fatigue which can cause a chain to break well below the ultimate strength of the material. These loads are due to the hydrodynamic currents in the water, aerodynamic loads on the pulling weight of the platform, and the local conditions causing the lines to have more or less sag depending on the load direction which can render the chains almost straight with a correspondingly higher horizontal tension component and stiffness, 'freezing' [9], hence fatigue. Microscopic physical damage accumulates with continued cyclic loading until cracks form. Once the crack reaches a critical size, brittle fracture occurs, the chain will break and the mooring will fail. A single mooring line failure may cause the platform to capsize. After multiple mooring failures the platform could drift away, losing control of the well-heads, which without de-pressurising would ultimately cause the risers to rupture catastrophically.

Several field studies have found that wear and tear occur in mooring chains links much sooner than anticipated, i.e. the combined wear and corrosion rate over the years is estimated to be 0.6 mm/year which is 50% higher than the maximum values found in corrosion inspection standards, e.g. API's RP2SK, DnV's OSE301. Loss of section of chain links could be due to corrosion, non-axial friction or even sulphate-reducing bacteria (SRB) that induces pitting corrosion, etc. (**Figure 2**). Consequently, there is an urgent need to either:

- increase the frequency of in-air testing, which would cause disturbance of operations at the platforms and decommissioning at each major inspection;

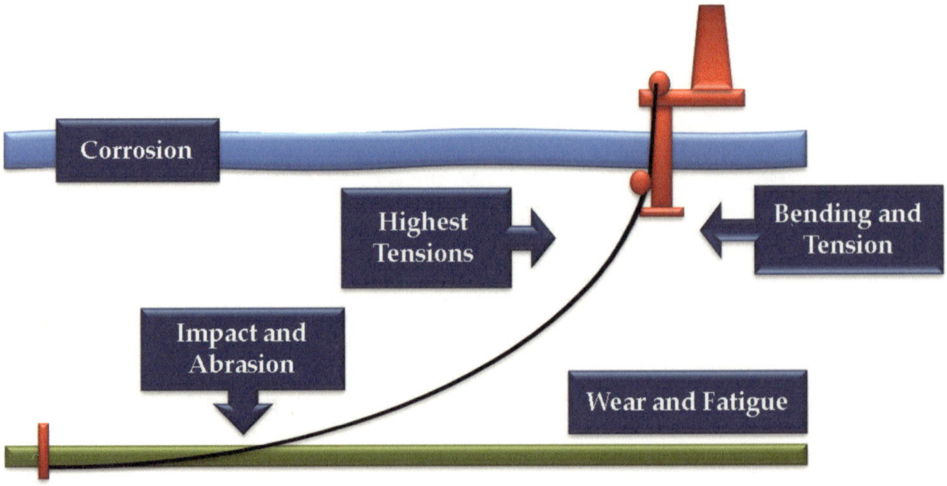

Figure 2. Mooring line degradation and the key areas to inspect.

- or increase the reliability of in-water NDT with a method that can assess progressive wear and tear while the facilities are in operation.

1.2. Innovative character of *in situ* monitoring in relation to the state of the art

Mooring chain life can be significantly reduced, leading to unacceptable risk of catastrophic failure, if early damage is not detected. Chain mounted equipment is available to monitor chain tension and bending, but detection of damage caused by stress concentrations, fatigue, corrosion and fretting or combinations of these is not currently possible. The acoustic emission (AE) technique is capable of detecting cracks in mooring chains and fatigue damage. AE monitoring has shown sensitivity to crack growth during fatigue tests on chains. This chapter will describe the AE technique for detecting fatigue cracks, a procedure for applying the technique, a methodology for incorporating the AE test data with other data in the frame of a holistic approach to integrity management of moorings and a specification for an operational system.

Structural health monitoring (SHM) is the process of implementing a damage detection and characterisation strategy for engineering structures. Damage is defined as changes to the material which adversely affect its performance. The extraction of damage-sensitive features from the very large amount of sensor data normally requires sophisticated statistical analyses.

2. State of the art of inspection methodologies

At present, the state of the art in-water inspection techniques are not reliable; experience has shown that anomalies identified by in-water inspection can only be evaluated with true confidence by in-air inspection.

Specifically, the in-water techniques do not provide early detection of fatigue cracks in the chains and consequently provide little early warning of loss of integrity of moorings. This is mainly due to the inherent difficulties in the logistics of underwater testing and to the inability of the techniques to reach all the areas within the chain links (i.e. contact surfaces between links, marine growth). Several in-water mooring chain NDT systems have been developed with varying levels of success. The main aim for all of them has been to reduce the level of 'human overlooking/presence' in water during the test. These range from a simple diver-deployed manual caliper to prototype stand-alone, ROV-deployed system and a chain climbing robot.

The in-water testing systems mainly deal with two inspection procedures:

- mooring chain system measurement management, i.e. to monitor change in dimensions from the manufacturer's set data as a sign of early development of fatigue cracks and corrosion. This method can be qualitative (provide an indirect measure of damage) or quantitative (provide a measurement);
- mooring chain system integrity, which is used for accurate component condition assessment. This is essential for real-time analysis of progressive defects and can act as an early alarm system.

2.1. Review of current practices

In principle, there are two major stages to the testing of mooring chains:

i. manufacturer's quality control inspection before deployment;

ii. in-service inspection, i.e. for safety and maintenance.

Both types use several known approaches:

- Invasive and destructive testing (IDT). This is usually carried out in-air and either before commissioning of a mooring system (i.e. sample testing) or after clear signs of early damage to the mooring chain (i.e. to establish causes of chain damage during its decommissioning of the chain for future chain design). The main IDT accredited checks are as follows:

  - break testing on at least three links of the same chain, e.g. an applied maximum load for a period of 30 seconds without showing signs of cracking;
  - mechanical testing (tensile and impact).

- Non-destructive testing (NDT) and visual inspection. These do not affect production and can be repeated. Besides visual inspection, the main types of NDT for mooring systems are as follows:

  - Magnetic particle testing (MPT);
  - Penetrant testing;
  - Radiographic testing;
  - Ultrasonic testing.

It is customary to recover the mooring lines part way through their service life for periodic in-air testing, but this has four disadvantages, namely:

(1) the lines may be damaged either during recovery or reinstallation, e.g. losing their studs;

(2) the whole operation is expensive, since the services of anchor handling and possibly heading control tugs will be required for a number of days;

(3) in-air inspection will not necessarily detect all possible cracks and defects;

(4) defects may grow between inspections.

The current situation in the water inspection of mooring lines is accurately reflected in the HSE UK Survey of in-water inspection:

> *'There is an imbalance between the critical nature of mooring systems and the attention HSE receive, i.e. embodied by the frequency and accuracy of real time testing. Currently, there is no in-water technique to check for possible fatigues, cracks and monitor the progressive cases of cracks and defects*

> *in a real time manner. A new inspection system is needed, which is mostly to be of acoustical nature… It is clearly not appropriate to rely on annual in-water ROVs inspection to check if a mooring line has failed' [4].*

From the above discussion, it is clear that an early detection tool for the structural condition of mooring chains would benefit operators to minimize the lost revenue related to unplanned shutdown of offshore oil and gas, wind platforms and other offshore platforms.

2.2. NDT procedures

NDT procedures are key documents. They state which technique is to be used (in NDT terminology, a technique is a specific way of applying an NDT method), the instructions on how it is to be used, including setting up the test equipment and its calibration, the data gathering processes and how the results are to be interpreted. The interpretation must include a methodology for sentencing test signals or indications and distinguishing them from spurious or non-relevant signals. All the NDT methods suffer from a propensity for giving false-calls, where defects are 'called' only to show when examined more closely that nothing is present. Many NDT techniques fall into disrepute when there are too many false calls and for this reason, special effort will be paid to developing procedures that are less prone to error.

The development of any NDT procedure starts with an understanding of the defects being sought. The most important influencing parameters on defect sensitivity in an NDT procedure for chains are the following:

- Type: the defects that are most likely to cause the chain link to fail are cracks emanating from the internal radius and caused by fatigue during fretting movement between the chain links during service and corrosion;
- Location: The internal radius is often the location of very small surface cold laps and cracks created during manufacture of the chain link and these can grow due to fretting between the interlocking chain surfaces. Other defects may be occur in the chain welds as a result of poor quality control during manufacture;
- Size: Chain-links are known with cracks up to half through-wall depth without failure. However, the criticality of cracks in chains is poorly understood. This presents a problem for NDT as there is a limit to the minimum detectable size. Moreover, if the test sensitivity is set too high, the test is likely to be slower and there will be more false-calls.

An example of a recent failure, investigation of a mooring chain link identified fretting in the contact area between the chain links (**Figure 3a**), and the propagation of one crack through the link thickness in a series of fracture faces of increasing diameter (**Figure 3b**).

In the following sections, two well-known SHM techniques have been put forward as an example of novel practices applied to this field: GUW and AE. In order to assess their application to mooring chain monitoring, both modelling and experimental methodologies and results will be described.

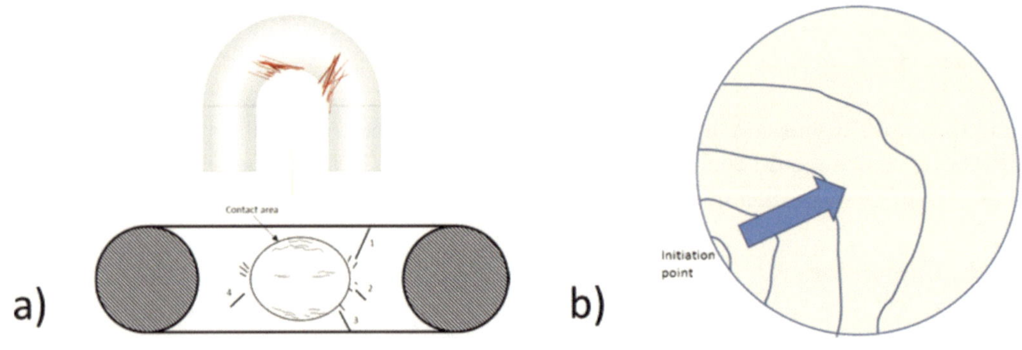

Figure 3. (a) Fretting of contact surfaces between chain links. (b) Fracture surfaces as crack propagates through chain.

3. Guided ultrasonic waves approach

A medium range ultrasonic test (MRUT) has been developed for chains that use guided ultrasonic waves (GUWs). GUWs propagate long distances along elongated objects such as pipes and cylinders, because the multiplying effects of internal reflections from the objects boundaries give rise to waves that are 'guided' and suffer relatively low energy losses. The wave modes are complex however. The so-called 'dispersion curves' (**Figure 4a**) show that as the frequency increases so does the number of wave modes. The additional wave modes increase 'noise' and have the potential to reduce test sensitivity. The high noise due to the presence of multiple GUW modes may be partly compensated with new signal processing algorithms that differentiate the higher-order modes. Alternatively, instead of relying on one ultrasound frequency in the test, the technique might involve a sweep through a range of test frequencies. Some experimental data have already been derived from chains in this way.

GUWs are used in the long-range ultrasonic testing (LRUT) of pipes. In LRUT, the transmitted wave mode from the transducer tool wrapped around the pipe is symmetrical and either longitudinal (L-wave) or torsional (T-wave). However, around chains, a symmetrical wave will become distorted by the chain curvature (**Figure 4b**) to become a flexural (F-wave). The distortion has been studied using numerical models supported by experimentation. Another option is to use Rayleigh waves instead of guided waves. These propagate along the surface only and exist at high frequencies when the frequency-thickness product is beyond a certain limit defined by the thickness of the pipe. However, Rayleigh waves are likely to be strongly affected by surface roughness.

3.1. Finite element modelling

Finite element analysis (FEA) has been used to study the complex GUW propagation around chains and therefore provide a theoretical basis for ultrasound frequency selection for chain links and to aid the optimisation of the inspection technique.

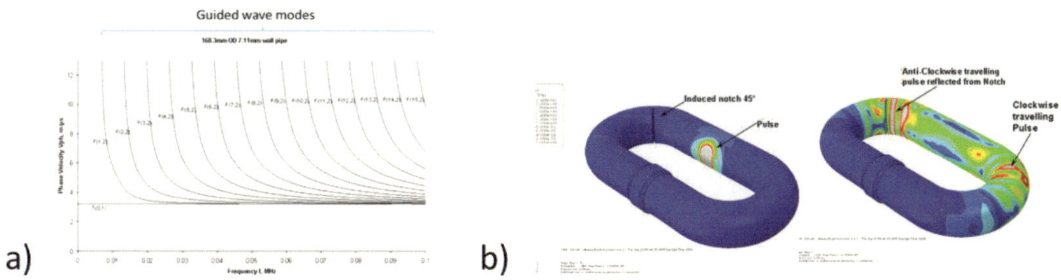

Figure 4. (a) Dispersion curves for a set of GUW modes. (b) Distortion of GUW around a chain.

The modelling work was conducting using the commercially available finite element software, Abaqus. The models were linear elastic and assumed the following material properties for carbon manganese steel.

- Young's modulus = 207 GPa
- Poisson's ratio = 0.3
- Density = 7830 kg/m^3

The finite element mesh was refined such that there were at least eight elements per wavelength for the smallest possible wavelength in the system. The elements used were eight-node linear bricks. In order to investigate the inspection of chain links, a number of models have been generated as follows.

- Natural frequency extraction models to calculate the dispersion characteristics of the straight section of the chain link. This modelling method [10] can be used with most commercial finite element software. It is able to calculate dispersion curves for prismatic structures of any cross section.
- Wave propagation models to calculate the mode conversion that occurs when GUW propagate around the bends in the link.
- Wave propagation analyses of the whole link including the weld at a range of frequencies.

A chain link of diameter 110 mm was used in the analysis.

3.1.1. Modelling results

The natural frequency analyses found that both the T(0,1) and T(0,2) exist at the typical torsional GUW inspection frequencies (20–80 kHz). **Figure 5** shows the displaced shapes and distribution of von-Mises stress in the straight section of the chain link at frequencies of around 45 kHz. The von-Mises stress has been used due to it being independent of the axis system used (e.g. Cartesian or cylindrical). It is proportional to the sound energy. **Figure 5** shows the distribution of von-Mises stress across the cross-section. It can be seen that the amplitude of the T(0,1) wave mode is strongest at the outside surface whereas the T(0,2) wave mode is

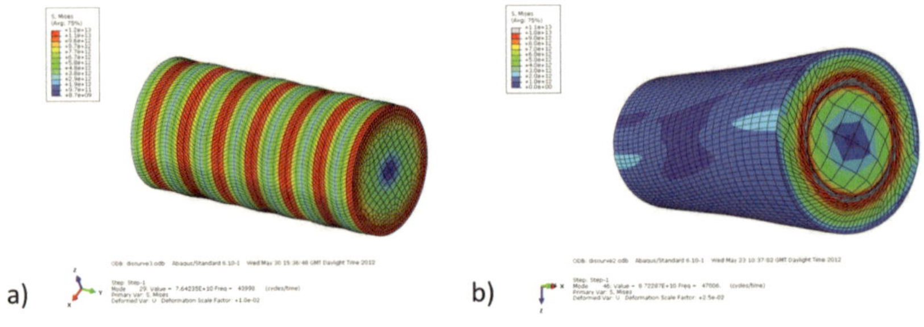

Figure 5. (a) Von-Mises stress distribution of the T(0,1) wave mode at 44 kHz and (b) T(0,2) wave mode at 47 kHz.

subsurface. The distribution of energy may affect the ability to detect flaws in a certain location. The results indicate that T(0,1) would be best for detecting surface breaking flaws. The natural frequency model was also used to extract the displaced shapes of torsional family wave modes.

Next, a wave propagation model was used to understand the behaviour of GUW as they propagate around the bend in the chain link. One bend was modelled and the ends of the model were elongated to prevent end reflections from interfering with the signals received. A single ring of exciters was used so that the pulse would propagate in both directions.

The magnitude of the displacement after excitation of a 10-cycle 40 kHz pulse is shown in **Figure 6**. It can be seen that the signal is no longer axisymmetric after propagation around the bend. This indicates that mode conversion has occurred. Some analysis was carried out to quantify the wave modes present in the signal after propagation around the bend. A mode filtering technique was used to separate the wave modes by circumferential order [12]. Since a torsional excitation was applied, it was assumed that wave modes in the torsional family were present. **Figure 6** shows the amplitudes of the individual wave modes plotted against circumferential order. It can be seen that there is a strong F(1,2) wave mode after passing the bend while T(0,1) wave mode propagates in the other direction along the straight section. The amplitude decreases with increasing circumferential order as would be expected.

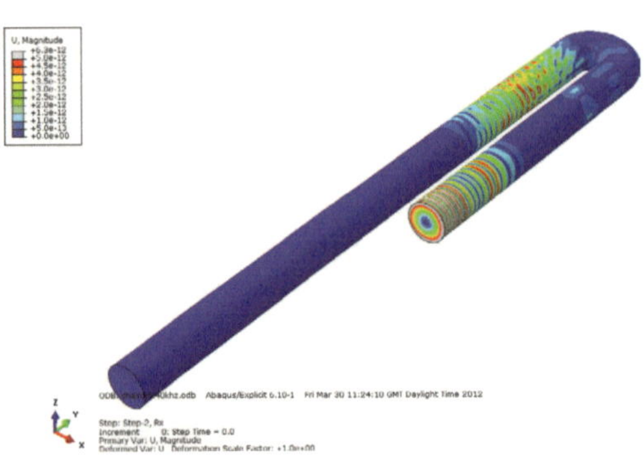

Figure 6. Displacement magnitude after propagation of a 10-cycle 40 kHz pulse around the bend.

Finally, a model of the whole chain link was created and a range of frequencies from 30 to 70 kHz were analysed. Excitation was applied using two rings to match the experimental work, where phasing is used to remove the wave propagating in one direction while reinforcing the wave propagating in the other. The ring spacing was 30 mm and 16 transducers around the circumference were simulated in each ring. The weld was idealised to a triangular shape with a height of 5 mm and a length of 60 mm on the opposite side of the chain from the ring. **Figure 7** shows the von-Mises stress in the chain link just after the input of a 10-cycle 30 kHz pulse. As before, it is clear that significant mode conversion has occurred.

Figure 8 shows the predicted A-scans from each of the models. The mode filtering technique was applied so that the A-scan for individual modes could be assessed. At 30 kHz, the reflections from the weld were distinct and there is relatively little 'noise' in between, whereas at 40 and 60 kHz, the reflections are less clear and the signals caused by the pulses of ultrasound circulating the chain become evident. The algorithm that is used to eliminate signals from pulses 'going the wrong' way through the rings starts to break down for certain wavelengths, and the circulating through-transmission pulses become superimposed on the pulse-echoes. At 50 kHz, there was a lot of noise at the start of the trace. This is likely to be caused by the T(0,2) wave mode. Its cut-off is around 50 kHz and therefore it is only excited at frequencies of 50 kHz and above. However, around its cut-off frequency, it will be highly dispersive which could cause this effect.

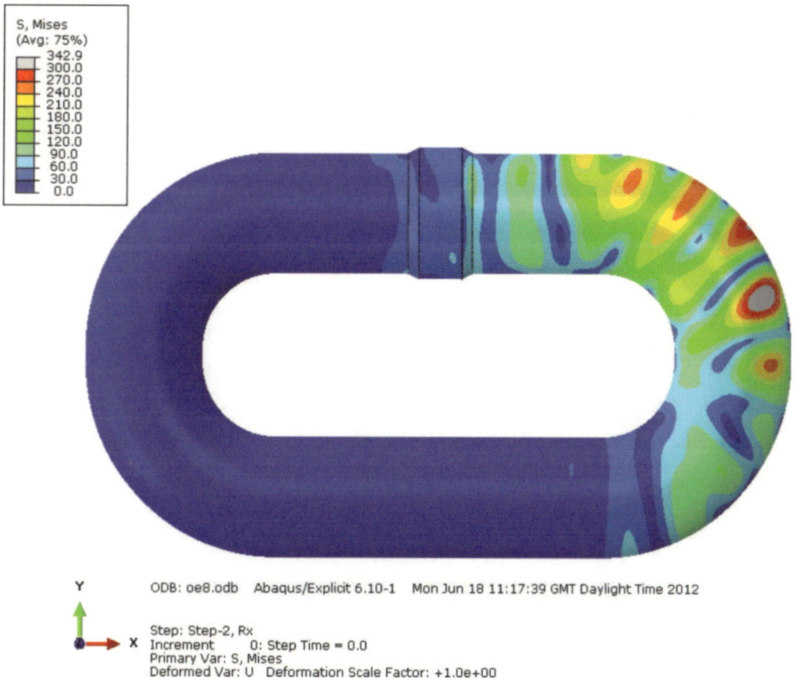

Figure 7. Von-Mises stress distribution in a chain link just after excitation of a 10-cycle 30 kHz pulse.

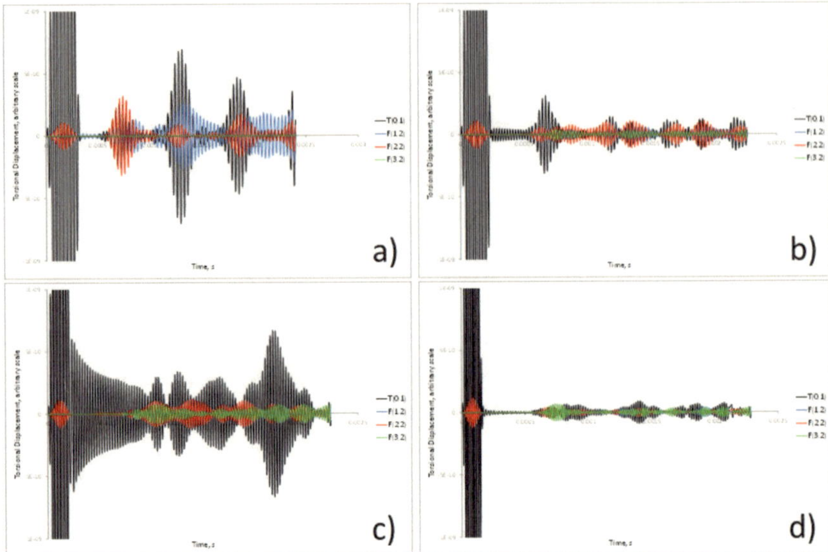

Figure 8. Predicted A-scan of individual wave modes for a: (a) 10-cycle 30 kHz excitation, (b) 10-cycle 40 kHz excitation, (c) 10-cycle 50 kHz excitation and (d) 10.

Finally, the model of the chain link was used to simulate a 50% cross-sectional area flaw for the 10-cycle 30 kHz case. The flaw was approximately 3 mm wide. **Figure 9** shows the layout of the model. **Figure 9** shows the predicted A-scans for individual modes. When compared with **Figure 10** it can be seen that the difference is quite noticeable indicating that detection of the presence of the 50% cross-sectional area of the flaw is possible.

3.2. Methodology and laboratory experiments on chain links

Modelling work was carried out, but under the important proviso that the test procedures used were not 'optimum'. In other words, further on-going work was/is needed on wave

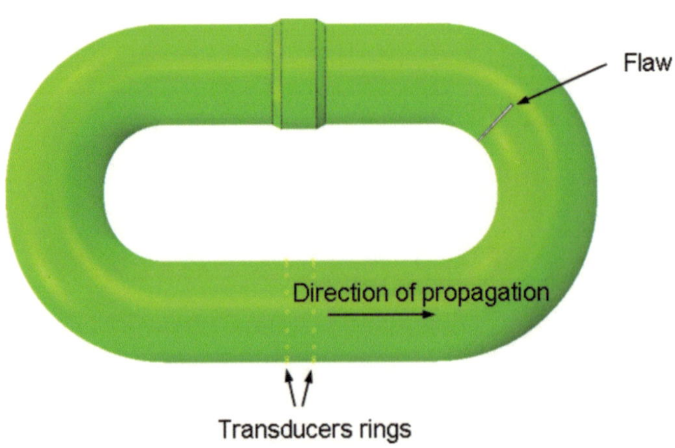

Figure 9. Layout of model of chain link with 50% cross-sectional area loss flaw.

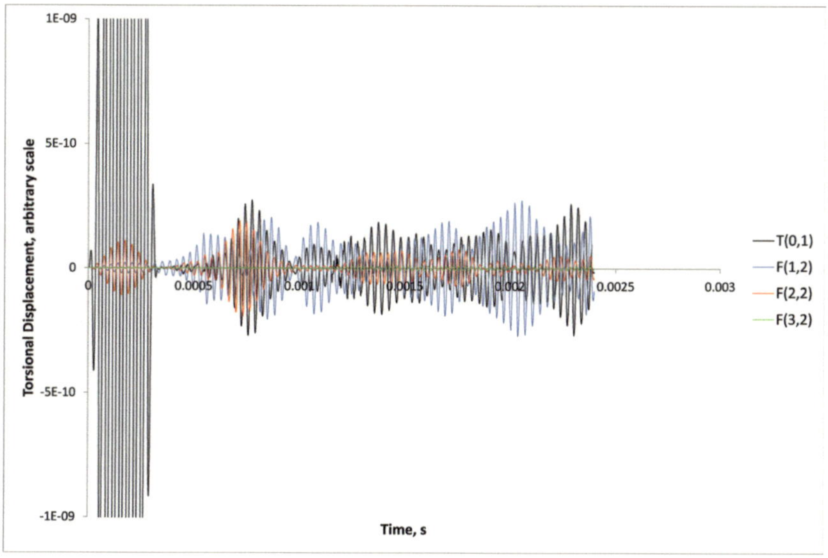

Figure 10. Predicted A-scan of individual wave modes for a 10-cycle 30 kHz excitation in a chain link with a 50% cross-sectional area loss flaw.

propagation in solid cylinders and around bends, selection of ultrasound frequency and pulse length, tool design, etc.

110 mm diameter chains were available for this work. Although this size is at the bottom of the range of chain sizes of interest to the end-users, the availability of the 110 mm diameter 3-m long solid rod on which to calibrate the A-scans was an important advantage at this stage of the study.

Eight chains were tested, two of which contained a defect by way of an EDM slot or a ground notch. The chains varied slightly from 105 to 110 mm diameter. The diameter is the diameter of the solid rod from which the chain link is forged. Once bent into shape it is welded at the ends to complete the whole link. The diameter determines the overall geometry of the chain.

Two slots were carefully placed on one of the 110-mm chain links; one on the intrados and the other on the extrados. On a second 105 mm chain, a notch on the intrados was 'grown' from 5 to 20% through-wall (**Figure 11**).

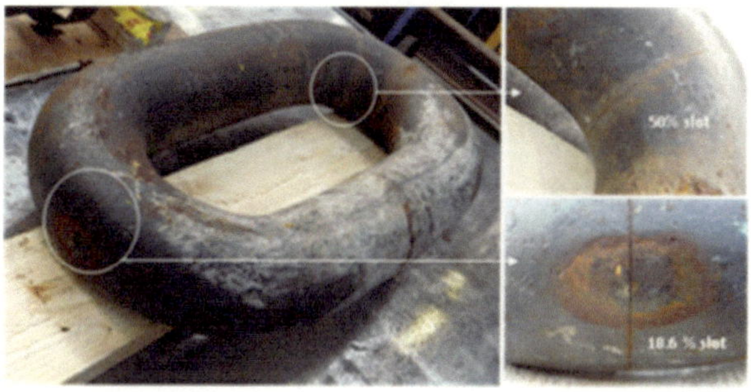

Figure 11. Aged chain link with slots.

3.2.1. *Guided ultrasonic waves equipment and specifications*

An important issue is whether the inspection capsule will be able to carry the instrument. There are advantages in keeping the distance between the instrument and the tool as short as possible, because noise is reduced and the signal is less attenuated.

In GUW techniques, the tool design and its performance is critical for the quality of the test results. To propagate symmetrical GUW into a pipe, the collar around the tool must apply equal pressure to all the transducers in a ring or an array.

In the present application, the transducer tool was always placed on the side of the chain opposite from the weld. Tests were performed with the transducer tool on the weld, but there was a drop in performance due to the unevenness of the surface.

The GUW frequencies were swept from 30 to 100 kHz in 5 kHz steps. The T-wave A-scans were collected over a 3-m range from the transducer and converted into ASC files for analysis.

3.2.2. *Experimental results*

A typical T-wave rectified A-scan is shown in **Figure 12**. It clearly shows multiple echoes from the weld on the opposite side of the chain form the tool.

The data could be grouped to show signal variation with frequency (**Figure 13**).

Closer analyses of the A-scans, however, show them to be divided into bands (see **Figure 14**):

(1) 30–40 kHz where the weld signals are clearly distinguishable;

(2) 50–60 kHz where a signal appears between the positions of the previous weld signals;

(3) 70–80 kHz where the weld signals are again clearly resolved;

(4) 95–100 kHz where signals cannot be distinguished from the noise.

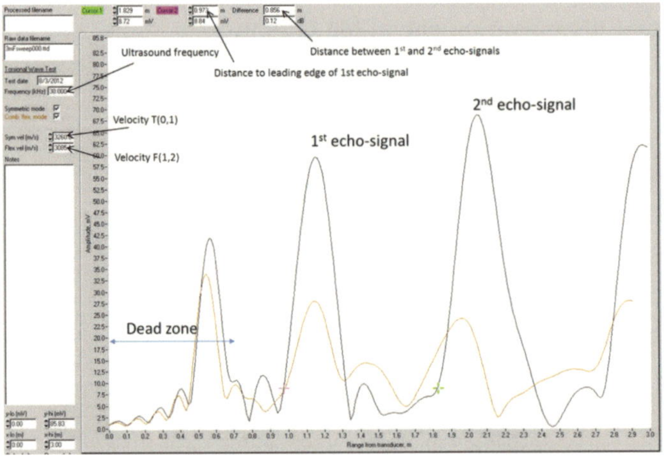

Figure 12. 30 kHz T-wave A-scan from 110 mm chain link.

http://dx.doi.org/10.5772/intechopen.68386

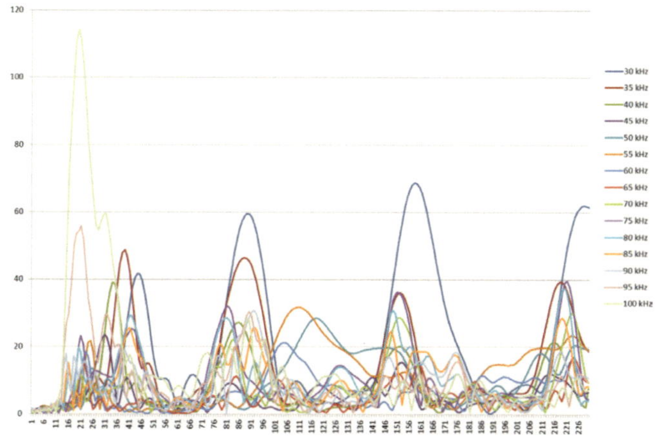

Figure 13. T-wave A-scan data collected from a defect-free chain: amplitude (mV) vs. time (ms).

These signal patterns were also observed in the finite element models. It is evident from the work here that only in the 30–40 and 70–80 kHz bands were the signals optimised. From the modelling, this appears to be a function of the ring spacing.

3.2.3. Conclusion

The propagation of GUW around mooring chains is extremely complex and the modelling and experiments reported here go only part of the way to explaining it. Nevertheless, T-wave propagation was proven to be sensitive to the large defects despite there being more wave modes than are present in LRUT of pipes. L-wave propagation along bars is extremely complex and only if signal processing methods can be developed to differentiate the modes might L-waves be considered for testing chains.

Also, the best resolution was obtained within certain frequency bands, evident in both the numerical modelling and the experiments.

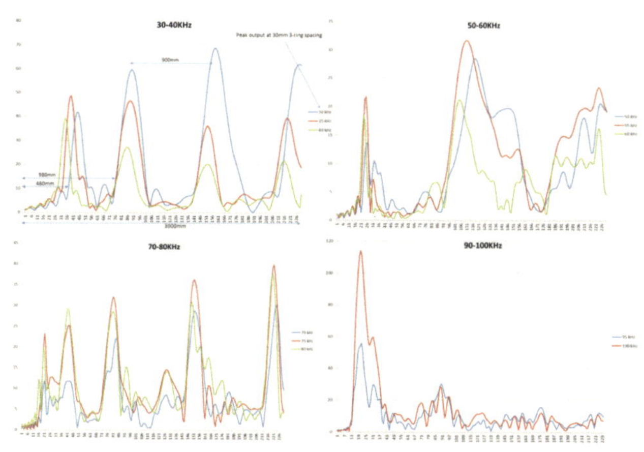

Figure 14. A-scans at different frequencies from 30 to 100 kHz: amplitude (mV) vs. time (ms).

4. Acoustic emission approach

Structural integrity approaches have strongly recommended monitoring mooring chains *in situ* during operation to verify mooring integrity. To more accurately assess the operational condition of in-service mooring chains, it is beneficial to investigate the next-generation of monitoring technologies and their ability to detect flaws and corrosion prior to critical failure. One promising monitoring tool for providing early warning of flaws is acoustic emission testing (AET), which has been used to successfully detect cracks in marine structures during operation.

Acoustic emissions are elastic waves that are spontaneously released by a material undergoing deformation. Acoustic emissions, or so-called 'hits' or events are the stress waves produced by the sudden internal stress redistribution of a material caused by changes in the internal structure. The stress can be hydrostatic, pneumatic, thermal or bending. Possible causes of the internal structure changes are crack initiation and growth, crack opening and closure, dislocation movements. Materials emit ultrasound when they are stressed and fail on a microscopic scale [11].

The optimum AE parameters must be estimated for each application. The appropriate selection and installation of the AE sensors is crucial for a precise data collection strategy. The data must be processed to determine crack initiation and growth and to discriminate irrelevant information.

Acoustic emissions are used to detect defects in structures both in service and during manufacture. The technique can also be used to monitor defect growth during mechanical test in the laboratory. It is an ideal method for examining the behaviour of materials deforming under load.

The difference between an AE technique and other NDT methods is the former detects active defects inside the material, while other the latter attempt to detect passive and active defects. Furthermore, AE needs only the input of one or more relatively small sensors on the surface of the structure or specimen being examined, so that the structure or specimen can be subjected to the in-service or laboratory operation, while the AE system continuously monitors the progressive damage.

The disadvantage of AE is that AE systems can only estimate qualitatively the extent of damage or size of defect. So, other NDT methods are still needed to do more exhaustive examination and provide quantitative results. Conventional ultrasonic evaluation is often used to evaluate AE indications.

4.1. Finite element modelling

Again FEA has been used to analyse the AE wave propagation along the structure. As described in section 3.1, the model is linear elastic and assumed the following material properties for carbon steel.

- Young's modulus = 207 GPa
- Poisson's ratio = 0.3
- Density = 7830 kg/m^3.

http://dx.doi.org/10.5772/intechopen.68386

A chain link of diameter 76 mm was used in the analysis.

A static analysis was run with a pressure of 1000 Pa to find the equilibrium state. The force was applied on the region shown **Figure 15a** which gave the result shown in **Figure 15b**

A dynamic model was created with the same geometry and same pressure applied, but with a crack inserted at the position indicated in **Figure 16**. The shape was a segment of the circle, with a maximum depth of 10 mm. The position was at the inner side of the join between the curved section and straight section of the chain link.

Stresses from the static model were applied to the dynamic model as the initial conditions.

Two AE sensors were modelled in the dynamic model. Each was 10 mm long, 29 mm around the circumference and positioned at 146.4 mm along from the plane of the crack. They were positioned one at the top of the model (Sensor 1) and one at the bottom (Sensor 2) as shown in **Figure 17**. The outputs were requested in the local cylindrical coordinate system (r, θ, z).

4.1.1. Modelling results

The dynamic model was solved in Abaqus for a simulated time of 0.5 ms. The crack opening can be observed in **Figure 18**.

The AE wave propagation and the displacement generated by the simulated crack growing can be observed in **Figure 19**. This relates directly with the elastic waves released at the crack tip.

The (r, θ, z) components of the displacements at each sensor location were recorded. **Figure 20** shows the displacement amplitude at both sensors location. The time of arrival (ToA) at each sensor can be observed. Calculating the value of the ToA, parameters such as the wave velocity or the location and time of occurrence can be estimated.

4.2. Methodology and experiments on chain links

Following the FEA analysis, a mooring chain link was monitored using AE in a simulated seawater tank tensile test rig. The rig is able to apply variable tension to a single link. A notch was initially introduced into the chain. During the test, a tensile load was applied at the

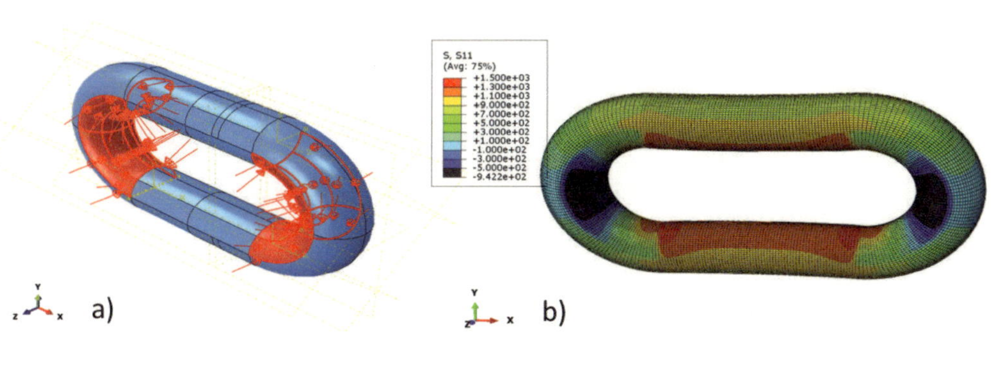

Figure 15. (a) Area were force is applied and (b) distribution of stress along the chain.

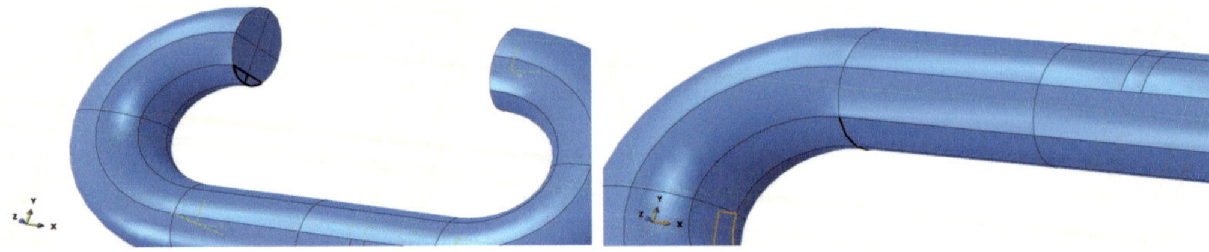

Figure 16. Chain seam model view.

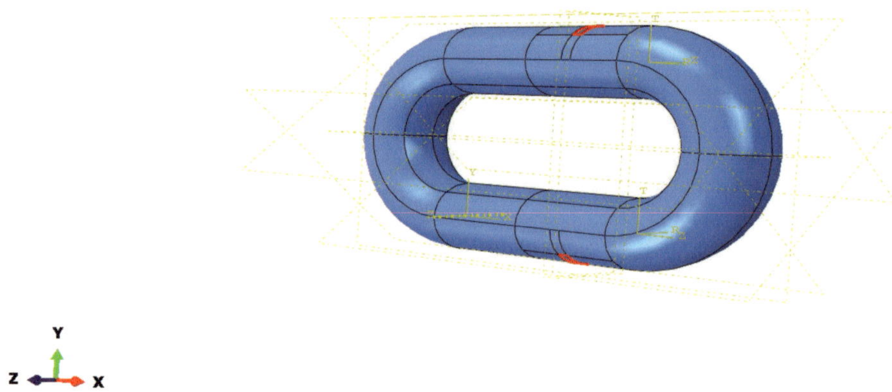

Figure 17. AE sensors location: Sensor 1 top, Sensor 2 bottom.

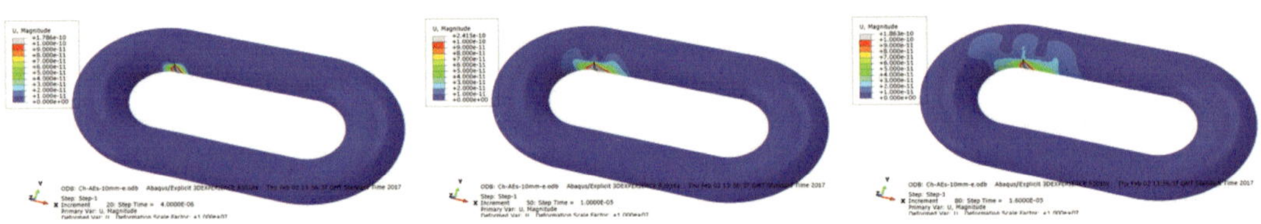

Figure 18. Crack opening model.

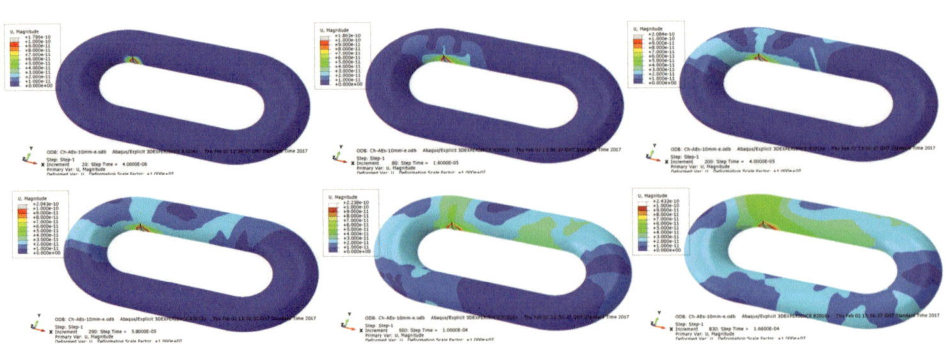

Figure 19. Elastic wave propagation model.

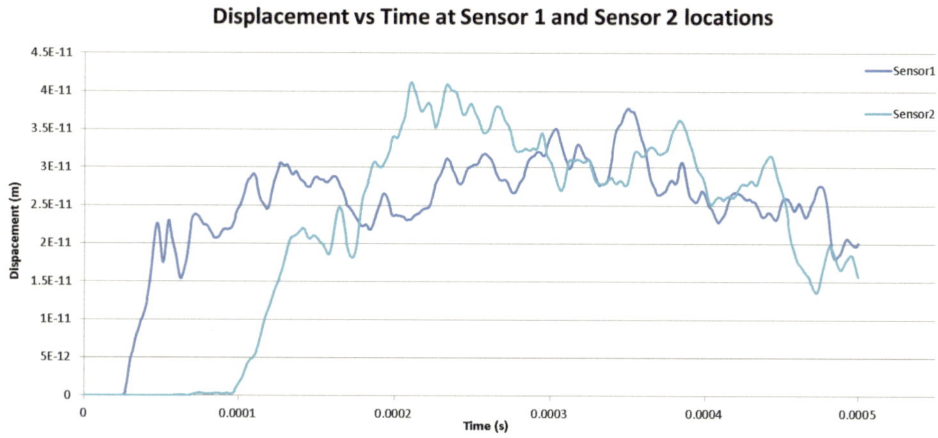

Figure 20. AE waveform for S1 and S2: displacement module vs. time.

connection point of the chain. The location of the sensors, forces and AE sensors can be seen in **Figure 21**.

4.2.1. Acoustic emission equipment and specifications

An AE system is a multi-channel data acquisition tool consisting of parallel measurement channels and system front-end software running on an external computer. A measurement channel consists of an AE sensor, AE preamplifier and one channel of an AE signal processor card.

The tensile test rig was instrumented with three AE sensors to record the AE data during the full test. Two AE sensors (150 kHz resonance frequency) were mounted with a magnetic holder on the tank frame to serve as 'guard' sensors in order to filter external environmental noise. The principal sensor used to detect the AE activities was submerged inside the seawater mounted using a magnetic holder. This was an AE sensor with an integrated preamplifier gain of 34 dB. Its resonance frequency was 150 kHz and it had an operating range of 90–450 kHz. It was of a design suitable for wet environments and on-site monitoring of underwater

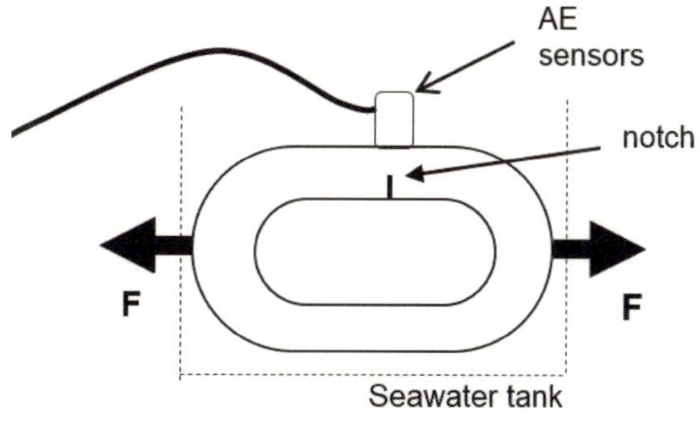

Figure 21. Chain set up under load with induced notch.

installations. The sensor was operated in a passive mode during the whole loading process to detect crack initiation and crack growth.

4.2.2. Experimental results

The laboratory trials were divided in two parts: the first part lasting for 75 hours to calibrate the set-up. In the second part, the system ran continuously for 285 hours to validate the long-term inspection capability. During these two tests, the load applied on the chain was kept constant at 8 MN.

One AE feature that proved to successfully represent crack initiation and propagation is energy. During the experimental test, cumulative energy was continuously calculated and recorded. From the calibration test, the set-up was validated (**Figure 22**).

During the long-term test, AE cumulative energy vs. time illustrated a linear increase in AE activity at first (**Figure 23**); this is followed by a rapid increase of energy when crack was propagating at a large scale.

4.2.3. Conclusion

AE graphs of cumulative energy vs. time show that the mooring chain crack propagation process was captured. The results can be considered as a characteristic curve of crack growth status over time.

Through both the calibration test and long-term test, the ability of the technique to detect and process AE events in real time has been proved. Other AE signal features including duration, peak amplitude together with cumulative energy should be analysed to evaluate the crack growth process.

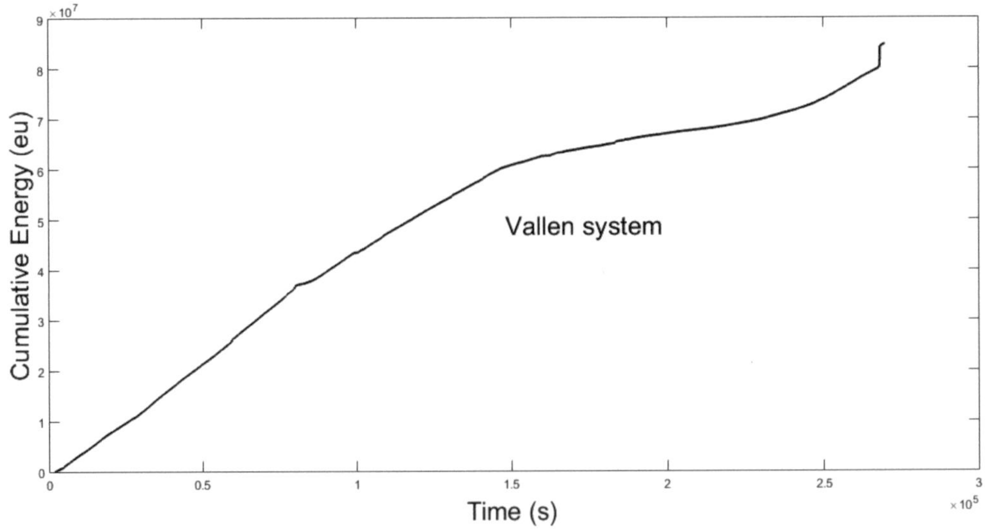

Figure 22. Cumulative energy vs. time (calibration test, 75 h).

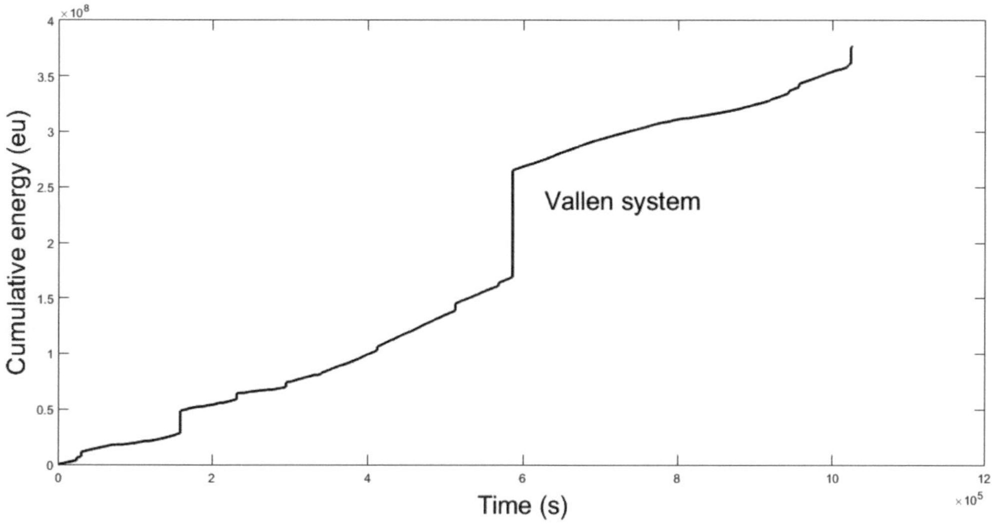

Figure 23. Cumulative energy vs. time (laboratory test, 285 h).

5. Final discussion

Due to the increasing demand of structural retrofit into conventional inspection strategies, SHM is of interest to an extensive range of industries. GUW and AE are non-destructive monitoring techniques which are widely employed at present. The output of its application will be comprehensive, real-time assessment of the structural condition of industrial assets.

The primary goal of this study was to investigate the applicability of GUW and AE approaches for crack initiation, location and propagation on a mooring chain. Modelling work and experimental testing have shown indication of the active damaged regions.

Because of the inherent uncertainties present in any SHM technique, the described technologies should be applied as part of a full mooring chain structural integrity assessment. Recent developments in internet infrastructure and connectivity for monitoring and sensing present an opportunity to overcome the limitations of AE and GUW testing for continuous monitoring. In addition to the continuous data output, a risk-based integrity management strategy may also include, where available, data from periodic inspections, numerical modelling showing stress distributions or crack propagation, historic and current operations.

Author details

Ángela Angulo, Graham Edwards, Slim Soua and Tat-Hean Gan*

*Address all correspondence to: tat-hean.gan@twi.co.uk

TWI Ltd, Integrity management Group, Granta Park, Great Abington, Cambridge, United Kingdom

References

[1] EU Commission Database on Number of FPS Drilling in European Waters. October 2010

[2] Applicable to the DnV Standard, International NDT (EN473 or ISO 9712)

[3] Lloyds Rules and Regulations for the Classification of a Floating Offshore Installation at a Fixed Location. Published by Lloyd's Register Group Limited, London, United Kingdom. Parts 1 to 11, 2013

[4] Floating production system. JIP FPS mooring integrity. Prepared by Noble Denton Europe Limited for the Health and Safety Executive, Norwich, United Kingdom, 2006

[5] International Energy Outlook 2016. Chapter 1. World energy demand and economic outlook. U.S. Energy Information Administration's, DOE/EIA-0484, May 11, 2016

[6] Richard Youngs. Energy Security, Europe's New Foreign Policy Challenge. Taylor & Francis, Abingdon, United Kindom, January 2009 Jan 28, 2009

[7] Farleyefndt M. Guidelines on the overall NDT quality system in Europe. In 17th World Conference on Non-destructive Testing, Shanghai, China; 2008

[8] Raj B. Non-Destructive Testing and Evaluation: Overview, Encyclopaedia of Materials: Science and Technology. India: Indira Gandhi Centre for Atomic Research; 2011. pp.6177-6184

[9] Larsen and Fylling, SeaTek Corp, Potts, Texas United States. 1982, 1981, 1993.

[10] Sanderson, R. M., & Smith, S. D.. The application of finite element modelling to guided wave testing systems. In D. O. Thompson, D. E. Chimenti, C. Nessa, S. Kallsen, & L. Poore, AIP Conference Proceedings (Vol. 657, No. 1, pp. 256–263). March 2003

[11] Huang, M, et al. Using acoustic emission in fatigue and fracture materials research. JOM. 1998;**50**:(11): 1-14

[12] Catton. Long range ultrasonic guided waves for the quantitative inspection of pipelines. Eng.D. thesis, School of Engineering and Design, Brunel University, London, England, 2009, pp. 2–4.

Chapter 6

Multi-Sensor Feature Extraction and Data Fusion Using ANFIS and 2D Wavelet Transform in Structural Health Monitoring

Ponciano Jorge Escamilla-Ambrosio, Xuefeng Liu,
Juan Manuel Ramírez-Cortés,
Abraham Rodríguez-Mota and
María del Pilar Gómez-Gil

Additional information is available at the end of the chapter

http://dx.doi.org/10.5772/intechopen.68147

Abstract

In this chapter, a novel feature extraction and data fusion approach for structural damage detection and localisation is presented. This approach combines adaptive network-based fuzzy inference systems (ANFIS) and two-dimensional wavelet transform (2D-WT) technologies. Simultaneous multi-sensor feature extraction and data fusion based on 2D-WT is carried out by forming a 2D multivariate signal, which is used to analyse the structure vibration response by measuring all sensors jointly. Energy values obtained from two-level db3 wavelet decomposition are arranged in a so-called energy percentage matrix (EPM), which is taken as an input for the ANFIS. The system is further trained by defining its output as the structural condition represented by a condition index. A set of output index patterns are defined depending on the level of damage assessment performed. The proposed method was tested through experiments using a cantilever beam structure. The testing results showed that the method is successful in detecting and localising damage by vibration analysis in structural health monitoring.

Keywords: sensor fusion, structural health monitoring, structural damage, ANFIS, 2D wavelet transform

1. Introduction

Structural health monitoring (SHM) refers to the process of damage detection, localization, quantification, or prediction, in infrastructure associated to fields such as aerospace, civil, or mechanical engineering. In engineering structures, damage is regarded as changes to physical and/or geometric properties of these systems which adversely affect their current or future performance [1, 2]. The interest in the ability to detect and locate structural damage at the earliest possible stage relies on safety issues with a consequence in economic benefits. Damage identification methods can be classified into four levels of damage assessment: level 1 (detection): determination that damage is present in the structure; level 2 (localisation): level 1 plus determination of the geometric location of the damage; level 3 (quantification): level 2 plus quantification of the severity of the damage; and level 4 (prediction): level 3 plus prediction of the remaining service life of the structure [3–5]. A variety of signal processing techniques have been attempted with good results in the field of SHM, such as statistical time series models, Fourier analysis, Wigner-Ville transform, Cohen class, Hilbert-Huang transform and different variations of wavelet analysis [6]. Data interpretation includes approaches such as neural networks, support vector machine, fuzzy and neurofuzzy schemes, Bayesian classifiers, or hybrid classifiers [7–11]. Among these, wavelet techniques (WT) have been proven to be an effective approach for damage detection, based on their characteristics to analyse non-stationary signals in both time and frequency domains. This analytical tool has gained preference in the research community to explore its applicability to perform structural damage assessment. Therefore, a variety of wavelet-based methods for damage detection have been proposed [12–14]. These methods can be broadly classified into three categories [15]: (1) variation of wavelet coefficients, (2) local perturbation of wavelet coefficients in a space domain and (3) reflected wave caused by local damage. The first category is generally used to find the existence and severity of damage. The second category is used to localise the damage in structures. The third category is used to measure the severity as well as the location of damage.

Under the premise that damage (e.g., cracks) in a structure will cause structural response perturbations at damage sites, a number of wavelet-based variants have been reported in the last decade in the field of SHM. In Ref. [16], an algorithm based on the wavelet packet transform and the Karhunen-Loéve transform is used to decompose the signals coming from an accelerometer on a vibrating composite beam. Wavelet packet transform was used as a feature extraction tool in SHM with good results. Integration of the discrete wavelet transform, an autoregressive model, damage-sensitive features, and support vector machine are proposed in Ref. [17] as a novel and efficient framework for damage detection of smart structures. Winkelman et al. [18] reported a novel signal-processing technique based on wavelet thresholding/denoising and Gabor wavelet transform. A damage identification method using unsupervised blind source separation with a wavelet-based pre-processing is presented in Ref. [19]. The authors report that independent component analysis biases to extract sparse components from the observed mixture signals, revealing damage instant and location.

In this chapter, an approach to structural damage detection and localisation (i.e. level 1 and level 2 damage assessments) based on the combination of two-dimensional WT (2D-WT) and adaptive network-based fuzzy inference system (ANFIS) techniques is described. The main novelty of the proposed approach resides in using 2D-WT to simultaneously perform multi-sensor data fusion and feature extraction for structural damage detection and localisation purposes. In addition, the use of an ANFIS system provides a fuzzy index damage representation for simultaneous identification and location of the damage.

http://dx.doi.org/10.5772/intechopen.68147

2. Background on 2D-WT and ANFIS

2.1. Wavelet transform (WT)

In the WT, a signal $f(t)$ is written as a series expansion in terms of wavelet families [20]:

$$f(t) = \sum_{k=-\infty}^{+\infty} < f(t), \varphi(t-k) > \varphi(t-k) + \sum_{j=-\infty}^{-1} \sum_{k=-\infty}^{+\infty} < f(t), \psi_{j,k}(t) > \psi_{j,k}(t) \tag{1}$$

where the father wavelet family $\{\varphi(t-k), k \in Z\}$ is used to describe the smooth part of $f(t)$ and the mother wavelet family $\{\psi_{j,k}(t), j \geq 0\}$ is used to describe the details of $f(t)$ at different levels. This expansion can expose the information originally hidden in $f(t)$.

An efficient way to implement the WT, and its inverse (IWT), is to use filter banks and downsampling/upsampling techniques. Moreover, the connection of WT with filter banks is a tool to understand the frequency allocation property of WT. **Figure 1** shows a two-level discrete wavelet decomposition and reconstruction, which demonstrates the idea of using filter banks to calculate WT and IWT.

The original signal $f(t)$ is broken down into three sub-signals: A_2, D_1, and D_2. From $f(t)$ to A_2, D_2, and D_1, the whole process could be seen as passing $f(t)$ through three filters (see **Figure 1b**). Each filter has different frequency characteristics and thus a frequency allocation is achieved through wavelet analysis. For example, the spectrum of the three filters associated with a two-level Haar wavelet analysis is shown in **Figure 2**. Note that the spectrum is plotted over the range [0, 0.5], where 0.5 corresponds to the Nyquist frequency (half of the sampling frequency).

The figure shows that the filter 1 acts as a low pass filter, the filter 2 serves as a band-pass filter and the filter 3 can be taken as a high-pass filter. Hence, by passing through these three filters, the original signal is split into three sub-signals and each sub-signal holds a different frequency content of the original signal $f(t)$: A_2 conserves most low-frequency content of $f(t)$, that is why it is called 'approximation'; D_1 contains the high-frequency content of $f(t)$, that is why it is called 'detail'; D_2 is another detail signal including the information of $f(t)$ which is not contained in A_2 and D_1. In this case, D_2 covers most of the middle-frequency content. These details and approximations at various levels may reveal valuable information of the signal characteristics that may not be clearly seen in the original signal $f(t)$.

2.2. Two-dimensional wavelet transform (2D-WT)

The two-dimensional (2D) wavelet representation is a straightforward generalisation of the one-dimensional (1D) wavelet representation. As in 1D, a 2D signal $f(x, y)$ can also be represented in terms of wavelet families. One difference between the 2D-WT in comparison with the 1D version is that all the signals in these wavelet families are 2D signals. In 1D wavelets, the mother wavelet family is generated by a basis wavelet function, i.e. a mother wavelet $\psi(t)$, and the father wavelet

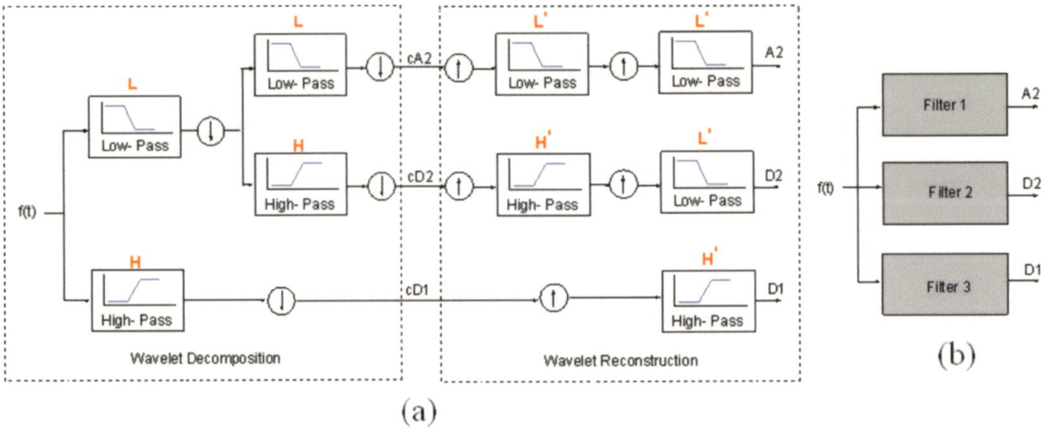

Figure 1. Filtering process of DWT and IDWT. (a) Two-level DWT decomposition and IDWT reconstruction using filter banks and (b) the three filters in (a).

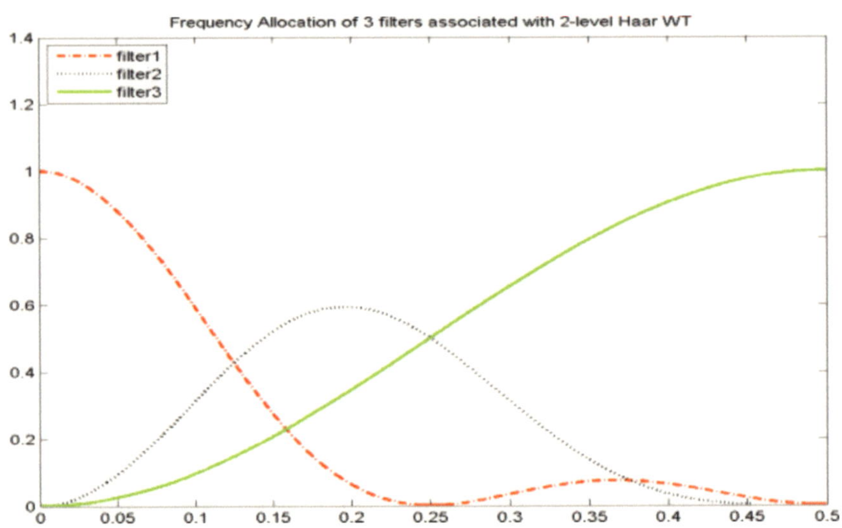

Figure 2. Frequency spectrum of the three filters associated with a two-level Haar wavelet analysis.

family is generated by another basis wavelet function, i.e. a father wavelet $\varphi(t)$. Similarly, in 2D-WT, wavelet families can also be generated from basis 2D wavelet functions [21, 22]. These basis 2D wavelet functions can be constructed by taking the tensor product (denoted as '⊗' in the following equations) of a horizontal basis 1D wavelet function and a vertical basis 1D wavelet function. This leads to four different types of 2D basis wavelet functions:

$$\Phi(x,y) = \varphi_h(x) \otimes \varphi_v(y) = \text{horizontal father} \otimes \text{vertical father} \tag{2a}$$

$$\Psi^v(x,y) = \psi_h(x) \otimes \varphi_v(y) = \text{horizontal mother} \otimes \text{vertical father} \tag{2b}$$

$$\Psi^h(x,y) = \varphi_h(x) \otimes \psi_v(y) = \text{horizontal father} \otimes \text{vertical mother} \tag{2c}$$

$$\Psi^d(x,y) = \psi_h(x) \otimes \psi_v(y) = \text{horizontal mother} \otimes \text{vertical mother} \tag{2d}$$

From Eq. (2a) to (2d), the basis 2D wavelet functions include one father wavelet and three mother wavelets. The corresponding four 2D wavelet families $\{\Phi_{m,n}(x,y)\}$, $\{\Psi^{v}_{j,m,n}(x,y), j\geq 0\}$, $\{\Psi^{h}_{j,m,n}(x,y), j\geq 0\}$ and $\{\Psi^{d}_{j,m,n}(x,y), j\geq 0\}$ are generated by scaling and translating these four-basis 2D wavelet functions as follows:

$$\Phi_{m,n}(x,y) = \Phi(x-m,\ y-n) \tag{3a}$$

$$\Psi^{v}_{j,m,n}(x,y) = 2^{-j}\Psi^{v}(2^{-j}x-m,\ 2^{-j}y-n) \tag{3b}$$

$$\Psi^{h}_{j,m,n}(x,y) = 2^{-j}\Psi^{h}(2^{-j}x-m,\ 2^{-j}y-n) \tag{3c}$$

$$\Psi^{d}_{j,m,n}(x,y) = 2^{-j}\Psi^{d}(2^{-j}x-m,\ 2^{-j}y-n) \tag{3d}$$

As with 1D wavelets, the father wavelet family is good at representing the smooth part and the mother wavelets are good at representing the details of a 2D signal. Hence, in 2D wavelets, the father wavelet family $\{\Phi_{m,n}(x,y)\}$ is used to describe the smooth part of $f(x,y)$, and three mother wavelet families, $\{\Psi^{v}_{j,m,n}(x,y), j\geq 0\}$, $\{\Psi^{h}_{j,m,n}(x,y), j\geq 0\}$, and $\{\Psi^{d}_{j,m,n}(x,y), j\geq 0\}$, are used to capture the vertical detail, the horizontal detail, and the diagonal detail of $f(x, y)$, respectively.

2.3. Adaptive network-based fuzzy inference system (ANFIS)

The combination of artificial neural networks (ANN) and fuzzy inference systems (FIS) has attracted the interest of researchers in various scientific and engineering areas due to the growing need of adaptive intelligent systems to solve real-world problems. ANNs learn by adjusting the interconnections between layers. FIS is a popular computing framework based on the concept of fuzzy set theory, fuzzy if-then rules, and fuzzy reasoning. There are several approaches that integrate ANN and FIS, and ANFIS is one of them. ANFIS can be treated as a FIS with a network-like structure or as an ANN containing fuzzy rules. Due to that, by adopting ANFIS in the form of an ANN clearly reveals its training ability, this form of treating ANFIS is explained as follows. For the explanation of ANFIS as a FIS, the interested reader is referred to Ref. [23]. For simplicity, it is assumed that a zero-order Sugeno fuzzy model has been adopted in the ANFIS structure. Hence, the ANFIS under consideration has two inputs x and y and one output z. Each input is assumed to have two fuzzy sets defined as their possible values, from the FIS point of view. Therefore, the ANFIS structure in this case can be seen as the five-layer ANN shown in **Figure 3**. The output of the ith node in layer j is denoted as O_i^j.

The function of each layer is as follows:

Layer 1: The function of nodes in the first layer is equivalent to the 'fuzzification' process in a fuzzy system.

$$O_i^1 = \mu_{A_i}(x), \text{ for } i = 1,2 \tag{4a}$$

$$O_i^1 = \mu_{B_{i-2}}(y), \text{ for } i = 3,4 \tag{4b}$$

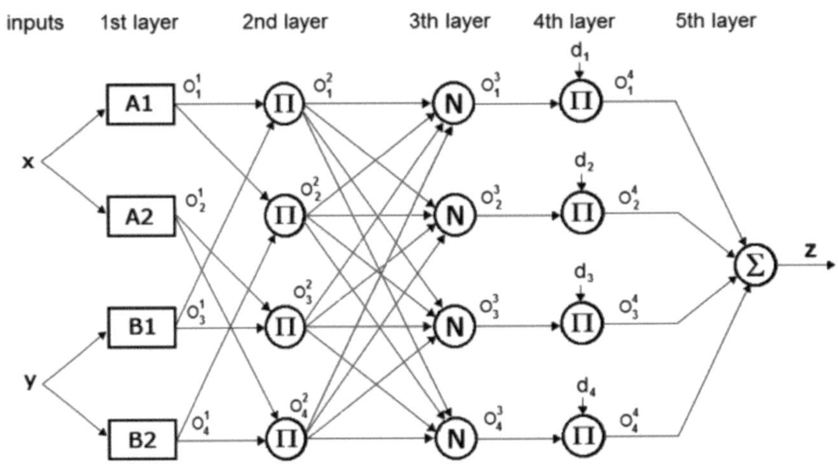

Figure 3. ANFIS architecture for a two-input Sugeno fuzzy model with four rules (from ANN point of view).

where μ_{A_i} and $\mu_{B_{i-2}}$ are the activation functions or membership functions (MF) defining the fuzzy sets for the inputs x and y, respectively.

Layer 2: Every node in this layer generates an output which is the product of all the incoming signals to that node. This process is equivalent to applying an 'AND' operator in the counterpart fuzzy system.

$$O_i^2 = O_1^1 O_{i+2}^1, \text{ for } i = 1, 2 \tag{5a}$$

$$O_i^2 = O_2^1 O_i^1, \text{ for } i = 3, 4 \tag{5b}$$

The output of this layer O_i^2 is the w_i, i^{th} firing strength, in the fuzzy system.

Layer 3: The ith node in this layer calculates the ratio between the output of the ith node in the previous layer and the sum of the outputs of all nodes in that layer:

$$O_i^3 = \frac{O_i^2}{\sum_{j=1}^{4} O_i^2} \tag{6}$$

The step in the counterpart fuzzy system is the normalisation of firing strengths.

Layer 4: Every node in this layer has a function of $O_i^4 = d_i O_i^3$, which applies the normalised firing strength (O_i^3) to the corresponding consequent parameter (d_i). The function of this layer corresponds to applying the implication method from a FIS point of view.

Layer 5: For the current ANFIS, there is only one node in this output layer. The node computes the overall ANFIS output through the summation of all incoming signals:

$$O^5 = \sum_{i=1}^{4} O_i^4 \tag{7}$$

The function of this node combines the steps of aggregation and defuzzification from the FIS point of view. Therefore, given a training set (input and output data pairs), ANFIS adjusts its parameters using well-developed learning algorithms. The parameters to be estimated in ANFIS are the premise parameters which define the membership functions in Layer 1, and the consequent parameters in Layer 4. ANFIS uses a hybrid-learning algorithm to estimate these parameters: a back-propagation learning algorithm is used to determine the premise parameters and least mean square estimation is adopted to determine the consequent parameters [23].

3. Multi-sensor data fusion and feature extraction using 2D-WT

The most common, as well as the most successful, application of 2D-WT is image compression. For instance, 2D-WT-based compression algorithms for transmitting and storing digitised fingerprints have been widely studied and used in the last decade [24, 25]. However, not much literature can be found of using the 2D-WT in structural health monitoring. In most applications, a set of many sensors are placed along different sections of a structure to measure its vibration response, accordingly the information from all the sensors needs to be integrated in some meaningful way. In this section, a novel method of using 2D-WT to carry out damage detection and localisation tasks is introduced. The 2D-WT provides an efficient and natural way to simultaneously extract features and integrate the information of many sensors, which makes the structural damage identification reliable and efficient.

The main idea of using the 2D-WT as a feature extraction and sensor data fusion mechanism is as follows: the structural response measured by one sensor i is a vector denoted as:

$$f_i(t) = [f_i(t_1) \quad f_i(t_2) \quad \dots \quad f_i(t_n)]^T \tag{8}$$

where n denotes the sampling number. Without losing any generality, it is assumed that l sensors (from sensor 1 to sensor l) are used to measure the vibration response of the structure under consideration. Thus, there are l measurement response vectors $\{f_1(t), f_2(t) \dots f_l(t)\}$. If these vectors are concatenated along rows and the result denoted as F, it gives:

$$F = [f_1(t) \quad f_2(t) \quad \dots \quad f_l(t)] \tag{9}$$

with F being an n-by-l matrix, which can be seen as an image, a 2D signal. A certain column of this image represents n vibration response samples measured by a sensor at some particular point of the structure, and a certain row of this image represents the vibration response samples measured by each sensor at l different location points of the structure at a particular sample time. The image F includes the information provided by all of the sensors throughout the measurement experiment duration and therefore gives a whole picture describing the dynamic behaviour of the structure under study. Therefore, if 2D-WT is applied to image

F, then its important features are revealed. A feature vector can be formulated in the following way: at the L-level 2D-WT, the original 2D signal (image) F is decomposed into 1 + 3L sub-2D signals (images): i.e. A_1, D_1^v, D_1^h, D_1^d, for a level 1 2D-WT decomposition.

The ratio of energy content of the 1 + 3L sub-signals in F is then defined as:

$$V = \left[\frac{\varepsilon_1}{\varepsilon_f}, \frac{\varepsilon_2}{\varepsilon_f} \cdots \frac{\varepsilon_{1+3L}}{\varepsilon_f}\right] \tag{10}$$

By arranging the 2D signal as a 1D signal, the terms $\varepsilon_1, \varepsilon_2, \ldots \varepsilon_{1+3L}$ in Eq. (10) are calculated using:

$$\varepsilon_x \overset{def}{=} \sum_{n=1}^{N} |x_n|^2 \tag{11}$$

which determines the energy of a discrete signal $x = [x_1, x_2, \ldots, x_N]^T$. The $\varepsilon_1, \varepsilon_2, \ldots, \varepsilon_{1+3L}$ terms represent the energy content in the sub-signals obtained from the 2D-WT. The term ε_f is the energy content in the original signal F. Each one of the components of V represents the energy percentage content of each of the 1 + 3L sub-signals in F. Therefore, a feature vector can be formed selecting some p components of V, which in turns would mean choosing some of the sub-2D signals as significant sub-signals contributing large energy percentages in F. The selection can be based on the following two criteria:

1. The sub-signals selected should be significant sub-signals contributing large energy percentages in F. The reason for this criterion lies in the fact that the insignificant sub-signals generated by WT are normally contributing to noise and should be removed. Empirically, it is assumed that a sub-signal is significant if its ratio of energy contribution to the original signal is no less than 3%.

2. The sub-signals selected should be sensitive to the damage. Damage usually has different effects on different frequency bands. For example, the effect of a crack on a beam is only noticeable within a high frequency band. Hence different sub-signals have different sensitivities to damage. By selecting the sub-signals sensitive to the damage, it is guaranteed that the damage could be effectively captured. The sensitivity analysis can be derived either by finite element modal analysis of the structure (analytical method) or by prior experiments (experimental method).

Although the two criteria above can be used, in this study, for convenience, we only choose the significant sub-signals, the ones with higher energy percentage content values, to form the feature vector.

Having chosen the energy percentage vector as the feature, the procedures for damage identification depend on the availability of the *a-priori* data. In an 'unsupervised learning mode', where data are only available from the undamaged structure, damage identification methods are based on feature comparison: two features, one extracted from the system in undamaged condition and the other from the current system, are compared in some way to obtain the damage indicator. The damage indicator is then compared to some threshold value and the conclusion about if the structure has deviated from the reference condition is obtained. On the other hand, in a 'supervised learning mode', where data from a system in different structural conditions (including the undamaged and some damaged conditions) are known in advance, the damage identification techniques are based on pattern classification: a database including models of the structure in different conditions is established using feature vectors for the *a-priori* data sets. Given a new data set which is to be classified as one of the conditions of the system, the task is to search through the database for the model which gives the best fit to the data. The corresponding condition of this database model is then applied to the data.

4. Structural damage detection combining 2D-WT and ANFIS

In this section, a detailed description on how to use the proposed ANFIS-2D-WT feature extraction method in structural damage identification is presented. As described before, the measurements of all the sensors are processed at the same time. Firstly, the measurements from all the l sensors in each test are arranged as a matrix and seen as a 2D signal (image), as it was explained in Section 3. Therefore, let us assume that the number of all possible conditions for the system is $r + 1$ (one healthy condition denoted as D_0 and r damaged conditions $D_1 : D_r$). It is assumed also, that for each condition, and considering all the sensors, a total of N images, or 2D data measurements are available. This results altogether in $(r+1)\times N$ image data sets that can be arranged as a data matrix:

$$Data\ Matrix\,2D = \begin{bmatrix} \{\alpha^{D_0\_1}\}, & \{\alpha^{D_0\_2}\}, & \dots & \{\alpha^{D_0\_N}\} \\ \{\alpha^{D_1\_1}\}, & \{\alpha^{D_1\_2}\}, & \dots & \{\alpha^{D_1\_N}\} \\ . & . & . & . \\ . & . & . & . \\ \{\alpha^{D_r\_1}\}, & \{\alpha^{D_r\_2}\}, & \dots & \{\alpha^{D_r\_N}\} \end{bmatrix} \tag{12}$$

where the 2D signals $\{\alpha^{D_j\_i}\}(i = 1\ldots N,\ j = 0\ldots r)$ correspond to output measurement data images from the ith test at condition D_j (2D signals, considered as images) from all the sensors. Each row of the matrix contains all the available image data sets for a certain condition.

Then, select a typical measurement 2D signal (or image) $\{\alpha\}$ in the data matrix and perform 2D-WT analysis on it. The original 2D signal $\{\alpha\}$ is decomposed into a number of sub-2D signals and the energy percentage contribution of each sub-2D signal into the original 2D signal is calculated. Based on the calculated energy percentages, p sub-signals are selected and the corresponding energy percentage feature vector is formulated. The energy percentages of the selected sub-signals are the inputs to the ANFIS model. Therefore, the number of input variables to the ANFIS is p. Three linguistic values characterised using linguistic terms as 'small', 'medium' and 'large' are defined for each of the p input variables. The type of these MFs for these linguistic values is 'bell-shape' and is defined by:

$$\mu_s(x) = \frac{1}{1 + |\frac{x-c_i}{a_i}|^{2b_i}} \tag{13}$$

Note that the linguistic values such as 'small', 'medium' or 'large' have different meanings for different input variables. There is only one output variable defined in the ANFIS: the structural health condition. For convenience, it is normally represented by a condition index. A zero-order Sugeno fuzzy model has been adopted in the ANFIS structure, which means that singleton values are defined for the output variable and the type of the corresponding MF is a distinct constant. So far, only the architecture of the ANFIS model has been determined: it contains p inputs (corresponding to p energy percentages) and one output (condition index). Each input variable has 'small', 'medium' and 'large' linguistic values characterized by three bell-shape MFs. The number of ANFIS rules is determined by the combination of linguistic values for the input variables. For p input variables, each with three linguistic values, the number of resultant combinations is 3^p. Correspondingly, the number of rules is 3^p. For example, assume that only two sub-signals are selected ($p = 2$). The two input variables and one output variable are denoted respectively as x_1, x_2 and z. For each input variable, three linguistic values denoted as $\{M_1^{x_1}, M_2^{x_1}, M_3^{x_1}\}$ (for x_1) and $\{M_1^{x_2}, M_2^{x_2}, M_3^{x_2}\}$ (for x_2) are defined. Therefore, a total of nine rules are contained in the ANFIS model:

Rule 1: If x_1 is $M_1^{x_1}$ *and* x_2 is $M_1^{x_2}$, then z is d_1

Rule 2: If x_1 is $M_1^{x_1}$ *and* x_2 is $M_2^{x_2}$, then z is d_2

Rule 3: If x_1 is $M_1^{x_1}$ *and* x_2 is $M_3^{x_2}$, then z is d_3

Rule 4: If x_1 is $M_2^{x_1}$ *and* x_2 is $M_1^{x_2}$, then z is d_4

....

Rule 9: If x_1 is $M_3^{x_1}$ and x_2 is $M_3^{x_2}$, then z is d_9

Once the ANFIS architecture and rules are determined, it is necessary to prepare data sets for training use. For each available output data $\{y^{D_j\_i}\}$ $(i = 1 \ldots N, j = 0 \ldots r)$ in the data matrix, perform the same 2D-WT analysis and the same p sub-signals are selected to form the feature vector. Their energy percentages are arranged as a vector denoted as $Per^{D_j\_i}$. This procedure is applied to all the data set in Eq. (12), and a matrix referred to as energy percentage matrix (EPM) is obtained:

$$EPM(WPT) = \begin{bmatrix} \{Per^{D_0\_1}\}, & \{Per^{D_0\_2}\}, & \ldots & \{Per^{D_0\_N}\} \\ \{Per^{D_1\_1}\}, & \{Per^{D_1\_2}\}, & \ldots & \{Per^{D_1\_N}\} \\ \cdot & \cdot & \cdot & \cdot \\ \cdot & \cdot & \cdot & \cdot \\ \{Per^{D_r\_1}\}, & \{Per^{D_r\_2}\}, & \ldots & \{Per^{D_r\_N}\} \end{bmatrix} \quad (14)$$

The vector $\{Per^{D_j\_i}\}$, containing p elements, is taken as an input vector for the ANFIS. EPM Matrix contains a total of $(r+1) \times N$ such input vectors for ANFIS. They are used as training data for ANFIS. The current ANFIS use a supervised learning algorithm, which means that the target output for each input vector is needed. It has been mentioned earlier that the ANFIS output is the structural condition represented by a condition index. Depending on the level of damage assessment conducted, different output index patterns are adopted. If the ANFIS is used only to identify damage occurrence (level 1 damage assessment), the output indices are Boolean values (0 for healthy condition D_0, 1 to r for damaged cases D_1~D_r). In this situation, the EPM data matrix can be appended to contain a total of $(r+1) \times N$ input and desired output data pairs:

$$Data\ Matrix(Level\ 1) = \begin{bmatrix} \{Per^{D_0\_1},\ 0\}, & \{Per^{D_0\_2},\ 0\}, & \dots & \{Per^{D_0\_N},\ 0\} \\ \{Per^{D_1\_1},\ 1\}, & \{Per^{D_1\_2},\ 1\}, & \dots & \{Per^{D_1\_N},\ 1\} \\ . & . & . & . \\ . & . & . & . \\ \{Per^{D_r\_1},\ 1\}, & \{Per^{D_r\_2},\ 1\}, & \dots & \{Per^{D_r\_N},\ 1\} \end{bmatrix} \quad (15)$$

If the ANFIS is used for the damage localisation (level 2 damage assessment), a total of $r + 1$ condition indices each corresponding to a structural condition need to be defined. Defining j as the index for condition D_j, the data matrix containing the input and desired output data pairs is:

$$Data\ Matrix(Level\ 2) = \begin{bmatrix} \{Per^{D_0\_1},\ 0\}, & \{Per^{D_0\_2},\ 0\}, & \dots & \{Per^{D_0\_N},\ 0\} \\ \{Per^{D_1\_1},\ 1\}, & \{Per^{D_1\_2},\ 1\}, & \dots & \{Per^{D_1\_N},\ 1\} \\ . & . & . & . \\ . & . & . & . \\ \{Per^{D_r\_1},\ r\}, & \{Per^{D_r\_2},\ r\}, & \dots & \{Per^{D_r\_N},\ r\} \end{bmatrix} \quad (16)$$

Data matrices Level 1 and Level 2 (Eqs. (15) and (16)) are used, respectively, for training ANFIS with two different levels of damage assessment.

The next step is ANFIS training. The number of the premise parameters to be determined is $3\times3\times p$. This comes from the fact that for each of the p input variables, three MFs, each defined by three premise parameters, were established. The number of the consequent parameters to be determined is 3^p. The ANFIS architecture uses a hybrid learning algorithm [23] to estimate these $9p + 3^p$ premise parameters all together with the consequent parameters.

Usually, in the implementation of a structural health monitoring system, a large number of sensors on the structure under analysis are strategically placed to provide an early indication of damage. The problem then is the processing of the signals collected from all the sensors. This problem is approached in this work by using the 2D-WT to process all the information provided by the sensors at once. This is possible by integrating all the signals into a matrix that can be considered as a 2D image. In this way the information provided by each sensor individually is taken into account, preserved and fused with the information from the other sensors.

ANFIS is used for the purpose of structural damage identification. However, there are various levels of damage assessments. Two ANFIS models, ANFIS1 and ANFIS2 are established accordingly. ANFIS1 is used to identify damage occurrence (level 1 damage assessment) while ANFIS2 is used for damage localisation (level 2 damage assessment). The architecture of these two ANFIS models is the same, but they produce different output values. ANFIS1 is only used to distinguish healthy and damaged conditions; therefore the output is a Boolean value (0 for healthy, 1 for damaged). The output of ANFIS2 is designed to differentiate all the possible conditions, which is performed by defining a numerical value j ($j = 0$ to 5), where j corresponds to damage condition D_j.

5. Experimental setup with a cantilever beam

For the sake of repeatability and insight, cantilevered beam experiment settings have been used by several researchers to demonstrate the feasibility of their proposed approaches to damage detection and localisation using a plethora of techniques, examples include Refs. [26–28]. Furthermore, several aerospace and civil structures such as rotor blades and bridges could be modelled as cantilever beams. Therefore, in this section the results of an experimental study involving shaker-excited vibration tests of an aluminium cantilever beam carried out in the laboratory are presented. The beam is 90 cm length and cross-section 2.545 × 0.647 cm. Zero-mean band-limited (0–500 Hz) Gaussian white noise was used as the input signal to the amplifier. The amplifier gain was controlled manually and the shaker provided an approximately 10 N peak, via a random force input to the beam. A force gauge screwed on the bottom surface of the beam was used to directly measure the input. The shaker was attached with this force transducer through a stinger. **Figure 4** shows the experimental setup.

Six accelerometers (7 g each) were screwed to the top surface along the centreline at selected positions (15, 30, 45, 60, 75, and 90 cm from the left fixed point, respectively). The data from each test came from these six accelerometers and one force transducer. The data were collected at a sampling rate of 10 kHz for a duration of 4 seconds. Five damage scenarios (*D*1 to *D*5) were simulated by adding a lumped mass (22 g) at 30, 45, 60, 75, and 90 cm, respectively (see **Figure 4**). A summary of the experimental damage conditions is provided in **Table 1**.

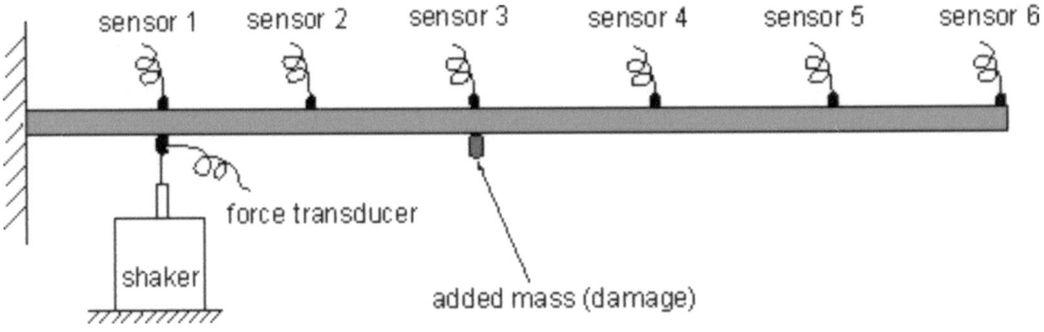

Figure 4. Cantilever beam experiment; setup schematic representation.

Damage case	Location of damage	Damage description
D1	0.30 m	Adding a lumped mass of 0.022 kg
D2	0.45 m	Adding a lumped mass of 0.022 kg
D3	0.60 m	Adding a lumped mass of 0.022 kg
D4	0.75 m	Adding a lumped mass of 0.022 kg
D5	0.90 m	Adding a lumped mass of 0.022 kg

Table 1. Summary of damage case D1–D5: adding a lumped mass at different locations.

6. Results

The experiment was repeatedly carried out under each of the six possible conditions of the system. The system response data and the corresponding condition were recorded during the test. From each condition, 20 test data are used, within which the first 10 are for the training use and the remaining 10 are for the testing use. Therefore, altogether there are 60 training data sets and 60 test data sets available. Thus, ANFIS is used for the purpose of structural damage identification. However, there are various levels of damage assessments. Two ANFIS models, ANFIS1 and ANFIS2 are established accordingly. ANFIS1 is used to identify damage occurrence (level 1 damage assessment) while ANFIS2 is used for damage localisation (level 2 damage assessment). The architecture of these two ANFIS models is the same, but they produce different output values. ANFIS1 is only used to distinguish healthy and damaged conditions; therefore the output is a Boolean value (0 for healthy, 1 for damaged). The output of ANFIS2 is designed to differentiate all the possible conditions, which is performed by defining a numerical value j (j = 0–5), where j corresponds to damage condition D_j.

As previously described, in the proposed ANFIS-2D-WT method the measurements from all six sensors in each test are arranged as a matrix. The two-level db3 2D-WT is performed on this matrix. From the obtained seven sub-signals, three are selected and their energy percentages are taken as inputs for ANFIS. Note that only one ANFIS1 model is needed for level 1 damage assessment and one ANFIS2 model is required for level 2 damage assessment. Each of these two ANFIS models includes the information of all six sensors. The mapping between the three inputs and one output of the trained ANFIS1 using ANFIS-2DWT method is shown in **Figure 5**. While **Figure 6** shows the results when applying the trained ANFIS1 to the testing data. In this figure, Exp means expected values, and Act means actual values. The testing error of ANFIS1 is 2.7567e-5. For the level 2 damage assessment, ANFIS-2D-WT also performs well. **Figure 7** shows the mapping relationship of the ANFIS2 model. The testing result of the trained ANFIS2 model is shown in **Figure 8**. We can see that the testing error of the ANFIS2 model is 0.0021.

The effect of noise is an influencing factor always present in the vibration response signals of structures. Therefore, to study the effect of noise on the proposed ANFIS-2D-WT method, random Gaussian white noise is added to the response of these 120 test cases. The noise intensity is defined by the signal-to-noise ratio (SNR):

$$SNR(dB) = 20\log_{10}\frac{A_{\text{signal}}}{A_{\text{noise}}} \tag{17}$$

where $A_{\text{signal}}$and A_{noise} refer to the root mean square (rms) amplitude of signal and noise, respectively. **Figures 9–11** show the testing results of ANFIS1 (level 1 damage assessment) with SNR = 20, 10, 5, respectively. The ratios of the RMS values between the noise and the signal for these three cases are 10, 31.6 and 56.2%, respectively. As can be seen in the figures, by using the ANFIS-2D-WT method, the damage can be correctly detected when SNR is no smaller than 10 dB.

Figures 12–14 show the testing results of ANFIS2 (level 2 damage assessment) at these noise levels. It can be seen that ANFIS2 can locate damage even when SNR = 10 dB. These

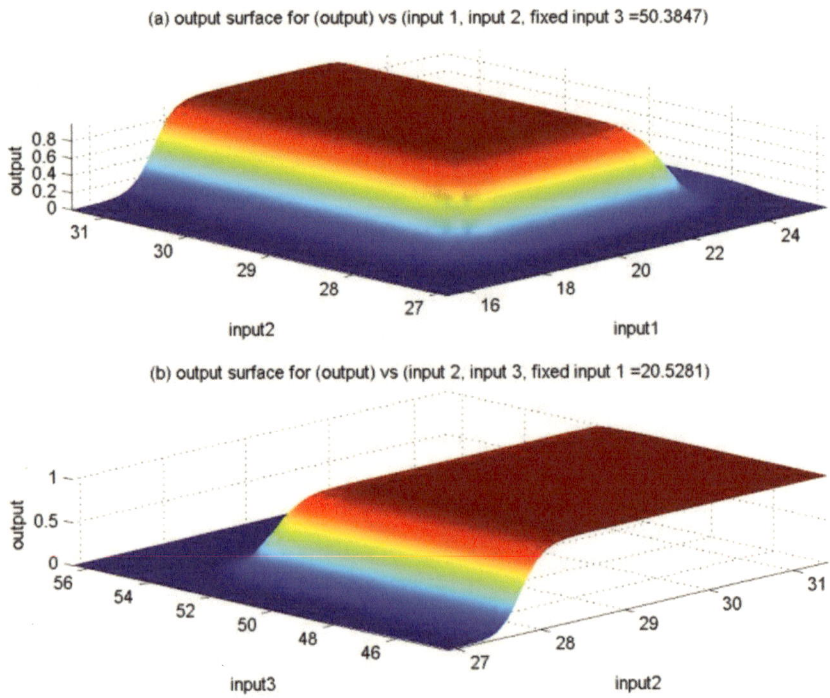

Figure 5. Output surface between ANFIS output and inputs: (a) input 3 is fixed to be 50.387 and (b) input 1 is fixed to be 20.5281 (ANFIS1, ANFIS-2D-WT method, using ALL six sensors).

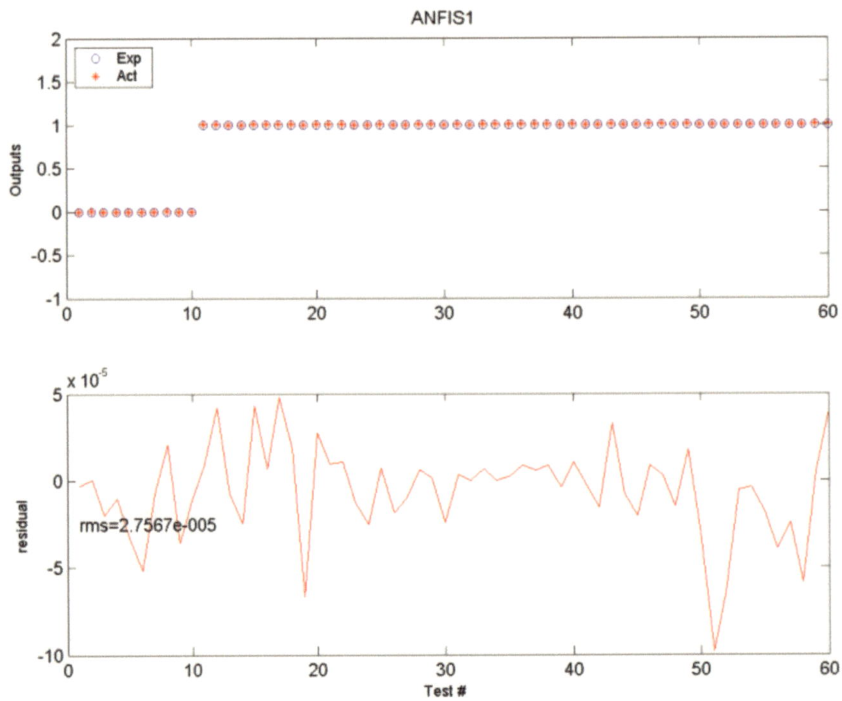

Figure 6. Testing results and the corresponding error curves (ANFIS1, ANFIS-2D-WT method, using all six sensors).

http://dx.doi.org/10.5772/intechopen.68147

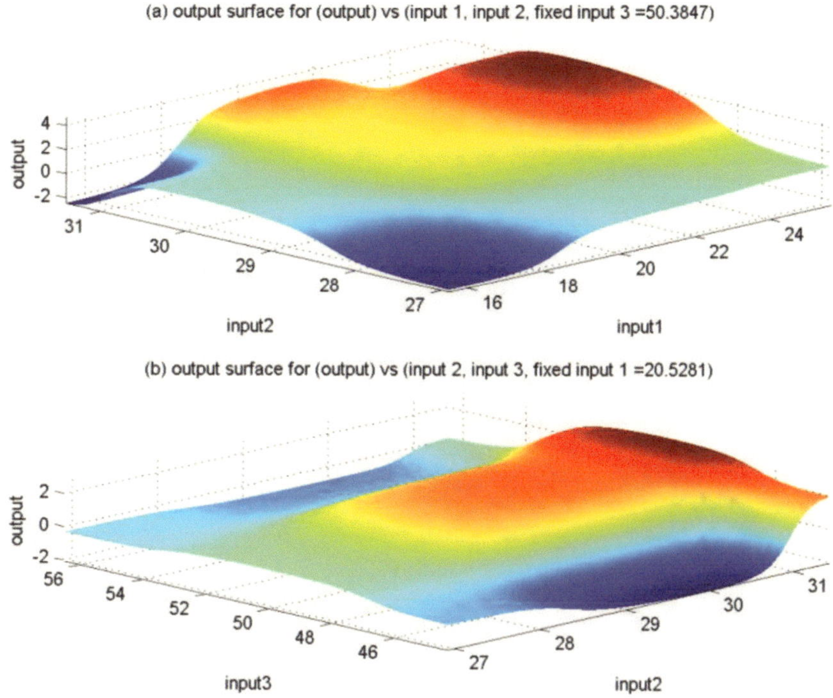

Figure 7. Output surface between ANFIS output and inputs: (a) input 3 is fixed to be 50.387 and (b) input 1 is fixed to be 20.5281 (ANFIS2, ANFIS-2D-WT method, using all six sensors).

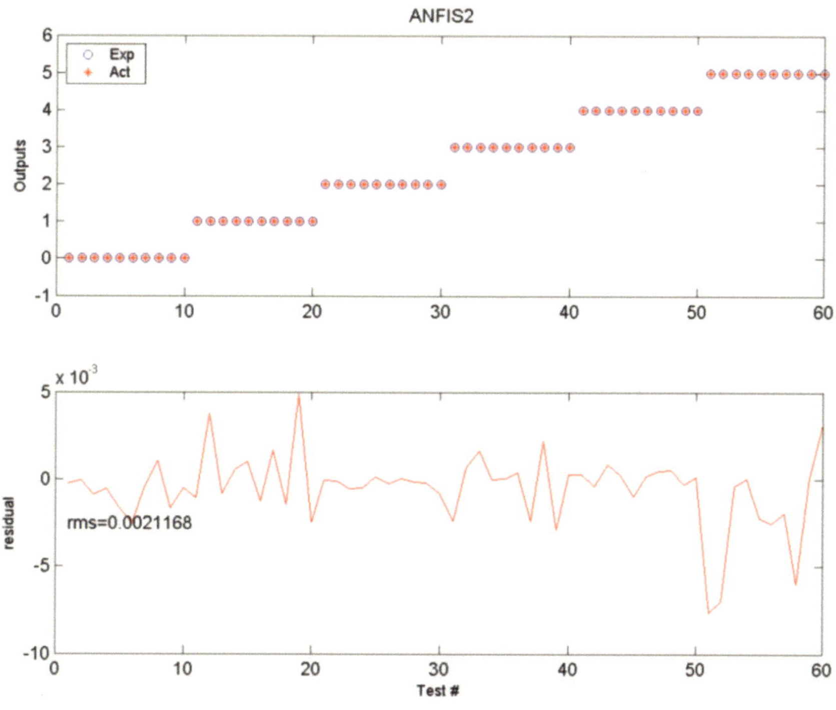

Figure 8. Testing results and the corresponding error curves (ANFIS2, ANFIS-2D-WT method, using all six sensors).

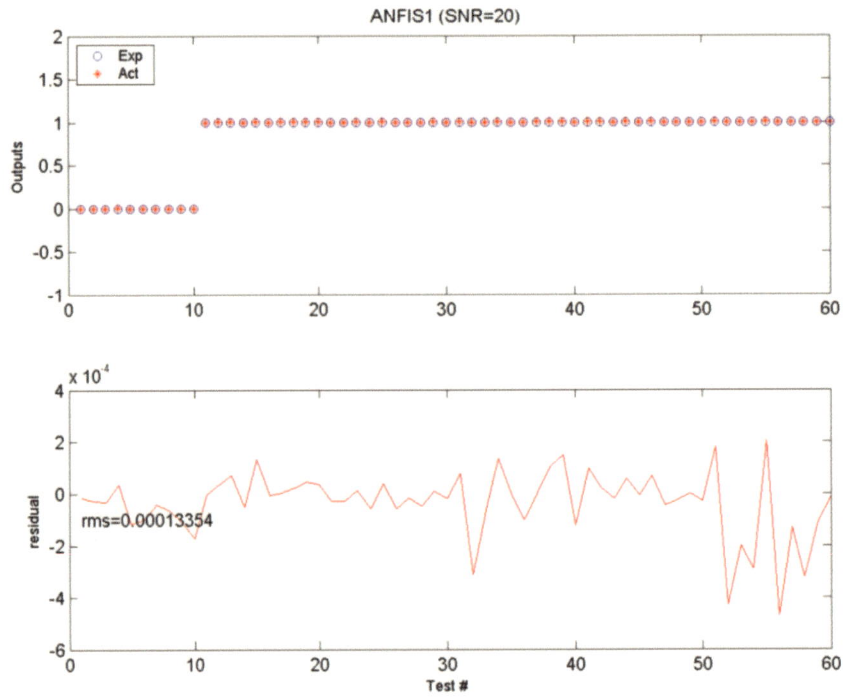

Figure 9. Testing results and the corresponding error curves (ANFIS1, ANFIS-2D-WT method, SNR = 20).

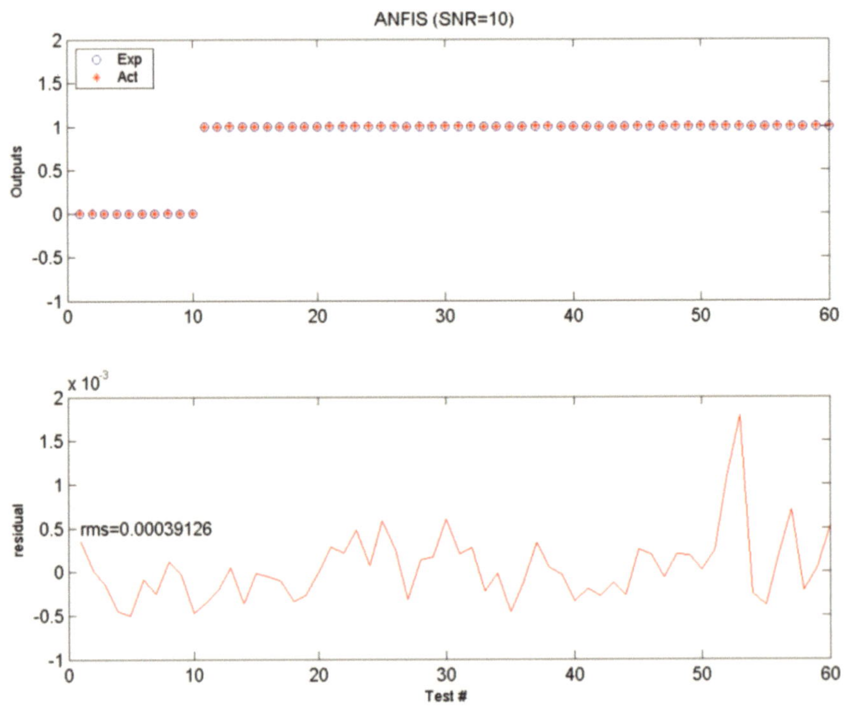

Figure 10. Testing results and the corresponding error curves (ANFIS1, ANFIS-2D-WT method, SNR = 10).

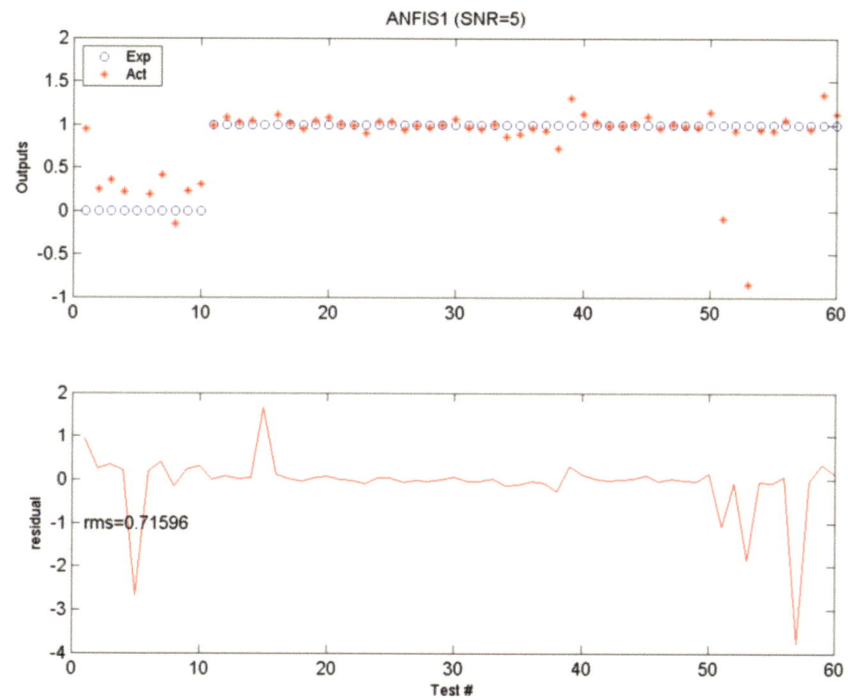

Figure 11. Testing results and the corresponding error curves (ANFIS1, ANFIS-2D-WT method, SNR = 5).

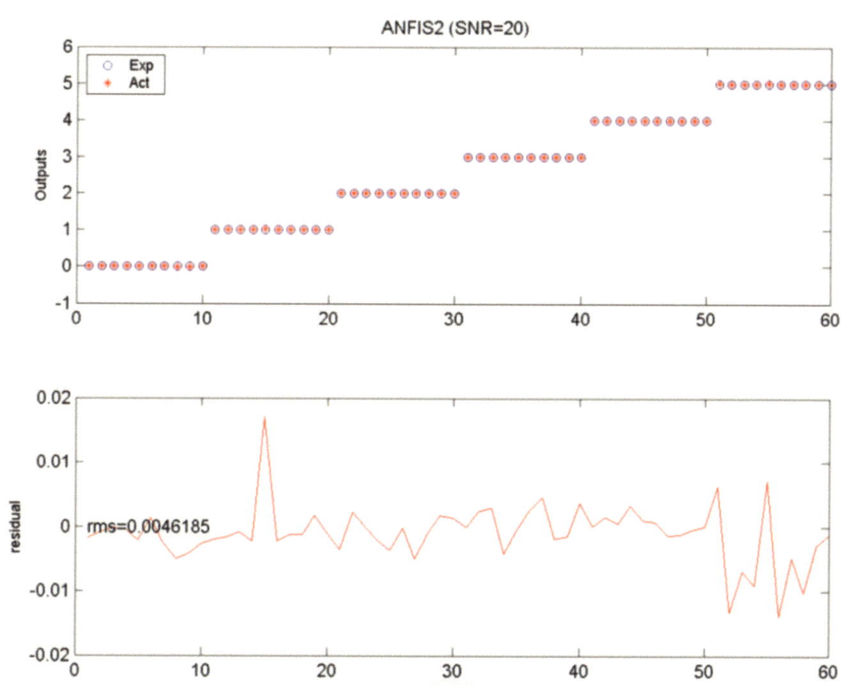

Figure 12. Testing results and the corresponding error curves (ANFIS2, ANFIS-2D-WT method, SNR = 20).

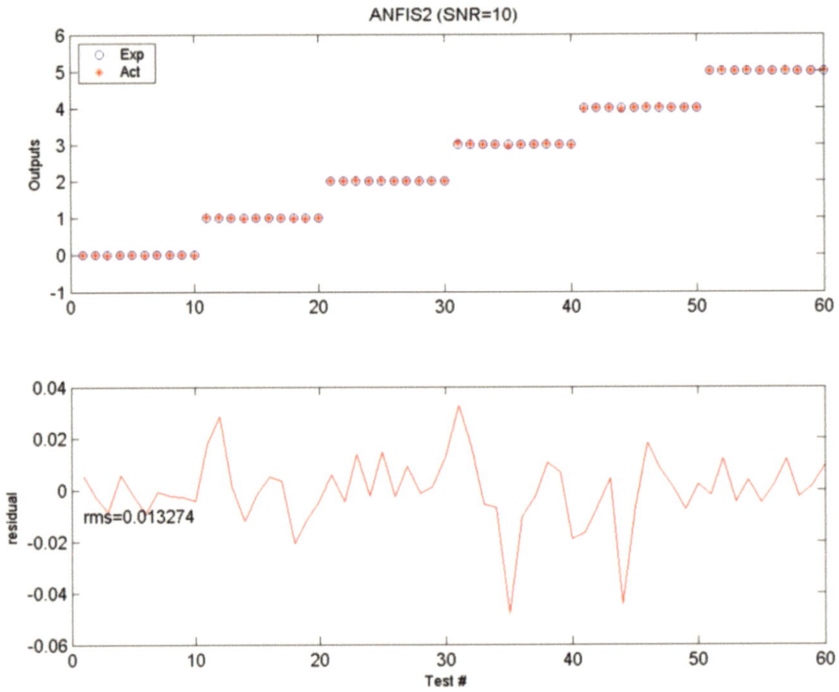

Figure 13. Testing results and the corresponding error curves (ANFIS2, ANFIS-2D-WT method, SNR = 10).

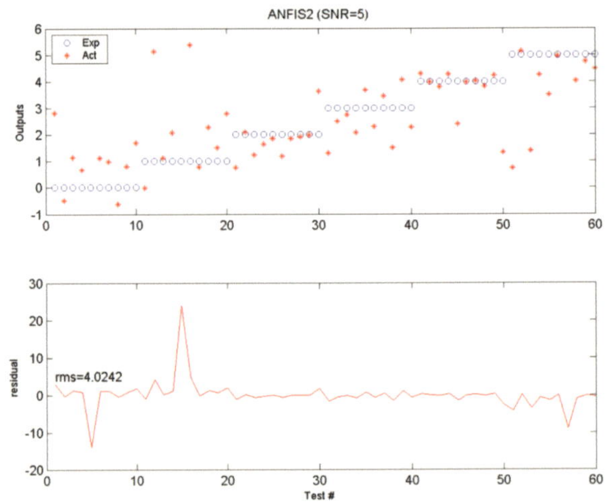

Figure 14. Testing results and the corresponding error curves (ANFIS2, ANFIS-2D-WT method, SNR = 5).

results suggest that measurement of noise does not seem to affect much of the proposed method for these two levels of damage assessment. This property can be attributed to the

http://dx.doi.org/10.5772/intechopen.68147

wavelet transform: the effect of noise can be alleviated by choosing sub-signals less affected by noise.

7. Conclusions

In this chapter, the ANFIS and 2D-WT technologies were combined to perform structural damage identification. The structure vibration response is decomposed by 2D-WT into a number of sub-signals, from which some are selected based on their energy percentages. The energy percentages of the selected signals are taken as inputs to the ANFIS model. The output of the ANFIS is a condition index, which can be a Boolean value (0 or 1) for level 1 damage assessment use, or a number of values for level 2 damage assessment use. Provided an ANFIS model is well-trained by the available data, it can be used for health monitoring and damage localisation. The proposed feature extraction method was applied to the data from a cantilever beam for damage detection and localisation. The testing results showed that the proposed method is successful in performing the two described levels of damage assessment. In addition, the ANFIS-2D-WT can process efficiently information from many sensors at the same time, performing simultaneously multi-sensor feature extraction and data fusion. The proposed damage assessment methodology of combining ANFIS with wavelet transform has great potential to implement systems which are able to interrogate sensor measurements autonomously for indications of structural damage.

Author details

Ponciano Jorge Escamilla-Ambrosio[1]*, Xuefeng Liu[2], Juan Manuel Ramírez-Cortés[3], Abraham Rodríguez-Mota[4] and María del Pilar Gómez-Gil[5]

*Address all correspondence to: pjorgeea@gmail.com

1 Instituto Politécnico Nacional, Centro de Investigación en Computación, México, Ciudad de México, Mexico

2 Department of Computing, The Hong Kong Polytechnic University, Hung Hom, Kowloon, Hong Kong

3 Department of Electronics, National Institute for Astrophysics Optics and Electronics, Tonantzintla, Puebla, Mexico

4 Instituto Politécnico Nacional, Escuela Superior de Ingeniería Mecánica y Eléctrica Unidad Zacatenco, Ciudad de México, Mexico

5 Department of Computer Science, National Institute for Astrophysics Optics and Electronics, Tonantzintla, Puebla, Mexico

References

[1] Farrar CR, Worden K. Structural Health Monitoring: A Machine Learning Perspective, West Sussex, UK: John Wiley and Sons; 2013.

[2] Balageas D, Fritzen CP, Güemes A. Structural Health Monitoring. London: ISTE Ltd; 2006.

[3] Srinivasan G, Massimo R, Sathyanaraya H. Computational Techniques for Structural Health Monitoring. London: Springer-Verlag; 2011.

[4] Sohn H, Farrar CR, Hemez FM, Shunk DD, Stinemates DW, Nadler BR. A review of structural health monitoring literature: 1996–2001, Los Alamos National Laboratory, Report LA-13976-MS, 2003.

[5] Santos JP, Cremona C, Orcesi AD, Silveira P. Multivariate statistical analysis for early damage detection. Engineering Structures. 2013;**56**:273–285

[6] Goyal D, Pabla BS. The vibration monitoring methods and signal processing techniques for structural health monitoring: A review. Archives of Computational Methods in Engineering. 2016;23(4):585–594.

[7] Amezquita-Sanchez JP, Adeli H. Signal processing techniques for vibration-based health monitoring of smart structures. Archives of Computational Methods in Engineering. 2016;23(1):1–15.

[8] Aminpour H, Foad N, Baghalian S. Applying artificial neural network and wavelet analysis for multiple cracks identification in beams. International Journal of Vehicle Noise and Vibration. 2012;**8**(1):51-59

[9] Wei F, Qiao P. Vibration-based damage identification methods: A review and comparative study. Structural Health Monitoring. 2011;**10**(1):83-111

[10] Ganguli R. Fuzzy cognitive maps for structural damage detection. Fuzzy Cognitive Maps for Applied Sciences and Engineering, Berlin, Heidelberg: Springer; 2014. pp. 267-290

[11] Arangio S, Bontempi F. Structural health monitoring of a cable-stayed bridge with Bayesian neural networks. Structure and Infrastructure Engineering, 2015;**11**(4):575-587

[12] Katunin A. Modal-based non-destructive damage assessment in composite structures using wavelet analysis: A review. International Journal of Composite Materials. 2013;**3**(6):1-9

[13] He WY, Songye Z. Progressive damage detection based on multi-scale wavelet finite element model: Numerical study. Computers & Structures. 2013;**125**:177-186

[14] Vafaei M, Alih SC, Baharuddin A, Rahman A, Adnan AB. A wavelet-based technique for damage quantification via mode shape decomposition. Structure and Infrastructure Engineering. 2014;**11**(7):869-883

[15] Kim H, Melhem H. Damage detection of structures by wavelet analysis. Engineering Structures. 2004;**26**:347-362

[16] Facchini G, Bernardini L, Atek S, Gaudenzi P. Use of the wavelet packet transform for pattern recognition in a structural health monitoring application. Journal of Intelligent Material Systems and Structures. 2015;**26**(12):1513-1529

[17] Kim Y, Chong JW, Chon KH, Kim J. Wavelet-based AR–SVM for health monitoring of smart structures. Smart Materials and Structures. 2013; **22**(1):015003

[18] Winkelmann C, Lestari W, La Saponara V. Composite structural health monitoring through use of embedded PZT sensors. Journal of Intelligent Material Systems and Structures. 2011;**22**(8):739-755

[19] Yang Y, Nagarajaiah S. Blind identification of damage in time-varying systems using independent component analysis with wavelet transform. Mechanical Systems and Signal Processing. 2014;**47**(1):3-20

[20] Mallat S. A Wavelet Tour of Signal Processing, The Sparse Way. 3rd ed. Burlington, MA: Academic Press; 2009.

[21] Antoine JP, Murenzi R, Vandergheynst P, Ali ST. Two-dimensional wavelets and their relatives. Cambridge: Cambridge University Press; 2004.

[22] Woods JW. Multidimensional Signal, Image, and Video Processing and Coding, Second Edition. Burlington, MA: Academic Press; 2011.

[23] Jang JSR, Sun CT, Mizutani E. Neuro-Fuzzy and Soft Computing: A Computational Approach to Learning and Machine Intelligence. New Jersey: Prentice-Hall; 1997.

[24] Shanavaz KT, Mythili P. Faster techniques to evolve wavelet coefficients for better fingerprint image compression. International Journal of Electronics, 2013;**100**(5):655-668

[25] Zhao S, Song, Wang XF. Fingerprint Image Compression Based on Directional Filter Banks and TCQ. In proceedings of Second International Workshop on Knowledge Discovery and Data Mining, Moscow, Russia; 2009. pp. 660-663

[26] Nataraju M, Adams DE, Rigas EJ. Nonlinear dynamical effect and observations in modeling and simulating damage evolution in a cantilevered beam. Structural Health Monitoring, 2005; **4**(5):259-282

[27] Kim DG, Lee SB. Structural damage identification of a cantilever beam using excitation force level control. Mechanical Systems and Signal Processing, 2010;**24**:1814-1830

[28] Chen H, Kurt M, Lee YS, McFarland DM, Bergman LA, Vakakis AF. Experimental system identification of the dynamics of a vibro-impact beam with a view towards structural health monitoring and damage detection. Mechanical Systems and Signal Processing. 2014;**46**(1):91-113